DATE DUE

GAYLORD			PRINTED IN U.S.A.

CHAOTIC DYNAMICS
OF SEA CLUTTER

Adaptive and Learning Systems for Signal Processing, Communications, and Control

Editor: Simon Haykin

Werbos / THE ROOTS OF BACKPROPAGATION: From Ordered Derivatives to Neural Networks and Political Forecasting

Krstić, Kanellakopoulos, and Kokotović / NONLINEAR AND ADAPTIVE CONTROL DESIGN

Nikias and Shao / SIGNAL PROCESSING WITH ALPHA-STABLE DISTRIBUTIONS AND APPLICATIONS

Diamantaras and Kung / PRINCIPAL COMPONENT NEURAL NETWORKS: THEORY AND APPLICATIONS

Tao and Kokotović / ADAPTIVE CONTROL OF SYSTEMS WITH ACTUATOR AND SENSOR NONLINEARITIES

Tsoukalas / FUZZY AND NEURAL APPROACHES IN ENGINEERING

Hrycej / NEUROCONTROL: TOWARDS AN INDUSTRIAL CONTROL METHODOLOGY

Beckerman / ADAPTIVE COOPERATIVE SYSTEMS

Cherkassky and Mulier / LEARNING FROM DATA: CONCEPTS, THEORY, AND METHODS

Passino and Burgess / STABILITY OF DISCRETE EVENT SYSTEMS

Sánchez-Peña and Sznaier / ROBUST SYSTEMS THEORY AND APPLICATIONS

Vapnik / STATISTICAL LEARNING THEORY

Haykin and Puthusserypady / CHAOTIC DYNAMICS OF SEA CLUTTER

CHAOTIC DYNAMICS
OF SEA CLUTTER

Simon Haykin

Sadasivan Puthusserypady

McMaster University

A WILEY-INTERSCIENCE PUBLICATION

JOHN WILEY & SONS, INC.

New York / Chichester / Weinheim / Brisbane / Singapore / Toronto

Copyright © 1999 by John Wiley & Sons, Inc. All rights reserved.

Published simultaneously in Canada.

For ordering and customer service, call 1-800-CALL WILEY.

Library of Congress Cataloging-in-Publication Data:
Haykin, Simon S., 1931–
    Chaotic dynamics of sea clutter / Simon Haykin and Sadasivan Puthusserypady.
        p.   cm.   — (Adaptive and learning systems for signal processing, communications, and control)
    Includes bibliographical references and index.
    ISBN 0-471-25242-5 (alk. paper)
    1. Radar—Interference—Mathematical models. 2. Backscattering—Mathematical models 3. Ocean waves—Remote sensing—Mathematical models. 4. Stochastic processes. I. Puthusserypady, Sadasivan.
    II. Title. III. Series.
TK6578.H39 1999
621.3848—dc21   98-51620
                                                                                         CIP

Printed in the United States of America
10 9 8 7 6 5 4 3 2 1

CONTENTS

Preface   **ix**

1 Overview   **1**

    1.1 Introduction / 1
    1.2 Objectives of the Book / 5
    List of Abbreviations / 5

2 Sea Clutter   **7**

    2.1 Introduction / 7
    2.2 Sea Clutter at Low Grazing Angles / 9
    2.3 Compound K-Distributions / 11

3 Radar and Database   **13**

    3.1 Introduction / 13
    3.2 IPIX Radar / 14
        3.2.1 System Features and Capabilities / 15
    3.3 Field Experiments / 17

4 Chaotic Dynamics of a Physical Time Series   **25**

    4.1 Introduction / 25
    4.2 Some Basic Definitions / 28
        4.2.1 Attractors / 28
        4.2.2 Basin of Attraction / 31
        4.2.3 Sink (Map) / 32
        4.2.4 Limit Cycle / 32
        4.2.5 Strange Attractor / 32

4.2.6   Manifold / 32
4.2.7   Fractal Dimension / 32
4.2.8   Lyapunov Spectrum / 35
4.2.9   Lyapunov (Kaplan–Yorke) Dimension / 41
4.2.10   Kolmogorov Entropy (KE) / 43
4.2.11   Relationship between the Kolmogorov Entropy
         and Lyapunov Exponents / 45
4.2.12   Embedding Dimension (d_E) / 45
4.2.13   Embedding Delay (τ) / 45
4.3   Criteria for Assessing the Chaotic Dynamics of an
      Experimental Time Series / 46

5   Preprocessing of the Radar Data   **47**

5.1   Introduction / 47
5.2   Amplitude and Phase Corrections (I–Q Calibration) / 48
      5.2.1   Effect of Miscalibration / 50
5.3   Filtering / 52
      5.3.1   Three-Point Smoothing / 52
      5.3.2   FIR Filtering / 54
5.4   Results of Preprocessing of the Signals / 56
5.5   Continuity and Differentiability Tests on Filtered Data / 58
5.6   Continuity / 59
      5.6.1   Definition / 59
      5.6.2   Test of Continuity / 60
5.7   Differentiability / 61
      5.7.1   Definition / 61
      5.7.2   Test of Differentiability / 62
5.8   Results of Continuity and Differentiability Tests / 64

6   Tests of Stationarity and Nonlinearity   **67**

6.1   Introduction / 67
6.2   Test of Stationarity / 68
      6.2.1   Recurrence Plots / 68
6.3   Tests of Nonlinearity / 71
      6.3.1   WSF Method / 72
      6.3.2   SIPD Method / 73
      6.3.3   SCD Method / 73
6.4   Results of Nonlinearity Tests / 74

7   Some Facts About Chaotic Processes   79

7.1   Introduction / 79
7.2   Sensitive Dependence on Initial Conditions / 81
7.3   Nonlinear Dynamical Systems / 85
7.4   What Is Phase Space? / 88
7.5   Why a Chaotic Approach? / 89

8   Reconstruction of Embedding Space   91

8.1   Introduction / 91
8.2   Measurements and State Representation / 91

   8.2.1   Embedding / 93
   8.2.2   Differentiable Embedding / 103
8.3   Phase-Space Reconstruction / 106
8.4   Estimation of Embedding Parameters / 108

   8.4.1   Embedding Delay / 108
   8.4.2   Global Embedding Dimension (d_E) / 110
   8.4.3   Local Embedding Dimension (d_L) / 111
8.5   Results of Estimation of Embedding Parameters / 113

   8.5.1   Comparison of d_E for Sea Clutter, Tides,
           and Correlated Noise / 114

9   Estimation of Chaotic Invariants   121

9.1   Introduction / 121
9.2   Estimation of Correlation Dimension / 121

   9.2.1   Maximum-Likelihood Estimate of the Correlation
           Dimension / 124
   9.2.2   Results of Case Studies / 126
9.3   Lyapunov Exponents / 130

   9.3.1   Estimation of the Lyapunov Exponents / 132
   9.3.2   Results of Case Studies / 133
9.4   Horizon of Predictability (HP) / 147
9.5   Lyapunov Dimension / 151

   9.5.1   Results of Case Studies / 152
9.6   Kolmogorov Entropy / 156

   9.6.1   How to Estimate the Kolmogorov Entropy / 157
   9.6.2   Maximum-Likelihood Estimation of
           the Entropy / 159
   9.6.3   Results of Case Studies / 161

10 Chaotic Study of Simulated Sea Clutter Data   167

10.1   Introduction / 167
10.2   Overview of the Simulation Model / 167

    10.2.1   Sea Surface Dynamics / 168
    10.2.2   Sea Surface Scattering / 173
    10.2.3   Radar Model / 178

10.3   Comparison Conditions and Criteria / 178

    10.3.1   Comparison Criteria / 178
    10.3.2   Determining Measurements Conditions and
            Simulation Parameters / 178
    10.3.3   Comparison of Measurement and Simulation
            Results / 179

10.4   Results of Chaotic Characterization / 179

11 Summary of Experimental Results and Conclusions   191

11.1   Supporting Physical Evidence / 193
11.2   Dynamic Reconstruction / 194

Appendix   195

    A.1   Estimation of Chaotic Invariants from Measurements of
        a Single Variable / 195
    A.1.1   Results / 196

Bibliography   201

Index   213

PREFACE

Radar backscatter from an ocean surface, commonly called sea clutter, has been traditionally viewed as the sample function of a stochastic process, not because the underlying physics of sea clutter says so but rather because of its random-looking waveform. The December 1997 issue of the journal *CHAOS*, published by the American Institute of Physics, included a paper entitled "Chaotic Dynamics of Sea Clutter" co-authored by the two authors of this book. In that paper we presented detailed experimental results that demonstrated, in the most convincing way for the first time, that sea clutter is in reality chaotic. This book is an expansion of that paper.

The interest of the senior author of this book in the chaotic dynamics of sea clutter dates back to the late 1980s. In particular, Henry Leung wrote a Ph.D. thesis under his supervision in which Henry claimed the chaotic dynamics of sea clutter by (1) demonstrating that the correlation dimension of sea clutter is fractal and (2) presenting some preliminary results on the dynamic reconstruction of sea clutter using a radial basis function (RBF) network. Those early results encouraged the senior author to probe more deeply into the chaotic dynamics of sea clutter throughout the 1990s, culminating in the co-authorship of this book.

This book should be of interest to radar engineers and scientists working on the characterization of sea clutter. It should also be of interest to researchers probing into the chaotic nature of experimental time series. The book presents the elements of a rigorous procedure for demonstrating whether the time series is chaotic or not. In that context, the experimental results presented herein on the chaotic dynamics of sea clutter could be viewed as a case study. An overview of the pertinent aspects of chaos theory is included in the book to make it essentially self-contained.

The authors wish to express their gratitude to the Natural Sciences and Engineering Research Council (NSERC), the Defence Research Establishment–Ottawa (DREO), and the Transportation Development Center (TDC) of Canada for financial support. They are indebted to their research colleagues Vytas Kezys and Brian Currie for their invaluable help in collecting the radar data used in this experimental study. The authors also wish to thank Henry D. I. Abarbanel, University of California, San Diego, Tasos Drosopoulos, DREO, Ontario, Tim Nohara, Sicom, Ontario, and Brian Currie and Vytas Kezys, both of McMaster University, for many useful comments on different parts of this book. The help of Brigitte Maier, Thode Library, McMaster University, in searching for and finding some difficult references is gratefully appreciated.

The use of the software tools RRCHAOS (developed at Delft University, The Netherlands) and cspX (developed by the Applied Chaos Group at Del Mar, California) for many of the computations performed in this book is gratefully acknowledged.

SIMON HAYKIN
SADASIVAN PUTHUSSERYPADY

Hamilton, Ontario
June 1999

CHAOTIC DYNAMICS
OF SEA CLUTTER

1

OVERVIEW

1.1 INTRODUCTION

At the very heart of classical physics was the assumption that the universe is strictly deterministic. Ever since Newton stated his three laws of motion, it had been taken for granted that any physical system, no matter how complex, must be deterministic because it can be described by a finite number of deterministic equations. For centuries, physicists firmly believed that to know the state of the system at any point in the future to arbitrary accuracy, all that was needed was a sufficiently accurate measurement of the state of the system at some prior time. The better the measurement, the better the prediction would be. Things changed at the beginning of the twentieth century, when physicists began to realize that the world is not completely deterministic. The revolution of quantum mechanics forced them to accept that some events in nature occur randomly and that there are fundamental limits on how accurately we can determine the state of anything. As bizarre as this discovery was, Newton's laws still held on a macroscopic scale, so, at least the everyday mechanical world still appeared safe from nondeterminism, but not for long. Physicists also found that even a simple system that obeys only Newton's laws can exhibit nondeterministic motion.

The emergence of nondeterministic motion from systems governed by purely deterministic laws is at first quite disturbing. The behavior of such systems is referred to as chaotic. It is extremely important to note from the beginning that to call such nondeterministic systems chaotic is very different from saying that it is random.

A chaotic phenomenon is capable of generating irregular and complex structures. Chaos implies unpredictability in time, while complexity implies irregularity in space. The spectacular development of chaos theory and fractal geometry has been made possible, on the one hand, by theoretical mathematical discoveries and, on the other hand, by the wide availability of powerful computers.

"Sea clutter," referring to radar backscatter from sea surface, has a long history of being modeled as a stochastic process that goes back to the early work of Goldstein [1951]. One of the main reasons for this approach has been the random-looking behavior of the sea clutter waveform. In the classical view, going back to Boltzmann, the irregular behavior of a physical process encountered in nature is believed to be due to the interaction of a large number of degrees of freedom in the system. This has led to the conclusion, in general, that complex behavior is due to the characterization of a complex system with a large number of free parameters, hence the justification for the stochastic approach. Contrary to this approach, recent developments in the theory of chaos and nonlinear dynamics have shown that many nonlinear and deterministic dynamical systems with relatively few degrees of freedom exhibit random/ complex/irregular behavior in their dynamics [Abarbanel, 1996; Abarbanel et al., 1993; Baker and Gollub, 1996; Lorenz, 1963; Mandelbrot, 1983; Moon, 1992; Ott, 1993; Peitgen et al., 1992; Schuster, 1988]. This, in turn, has opened up a powerful and elegant way of looking at the complex behavior of physical processes whereby they can be modeled as simple, nonlinear, and deterministic dynamical systems. Here, the word *simple* means a system with few degrees of freedom. We are reminded here of the fact that one of the fundamental issues in the mathematical modeling of a physical process is that of minimizing the number of degrees of freedom required to describe the process [Abarbanel, 1996; Abarbanel et al., 1993; Baker and Gollub, 1996; Lorenz, 1963; Mandelbrot, 1983; Moon, 1992; Ott, 1993; Schuster, 1988].

With the emergence of powerful analytic methods for handling time series generated by nonlinear dynamics, a different set of descriptive measures, for sea clutter is now available. Indeed, recent experimental results suggest that there may be a low-dimensional dynamical attractor controlling the behavior of sea clutter [Haykin, 1992, 1995]. Haykin and Leung were the first to claim the chaotic dynamics of sea clutter by looking at the fractal correlation dimension (D_2) [Haykin and Leung, 1989, 1992; Leung and Haykin, 1990]. They used the Grassberger–Procaccia [1983a & b] algorithm (GPA) to estimate (D_2), and its value was reported to be between 6 and 9 [Haykin and Leung, 1989, 1992; Leung and Haykin, 1990]. In 1992, He and Haykin moved one step forward to verify the chaotic dynamics of sea clutter by calculating the largest Lyapunov exponent. They used Wolf's method [Wolf et al., 1985] and showed that the largest Lyapunov exponent is indeed positive. Later, Haykin and Li reestablished the presence of a low-dimensional chaotic attractor for sea clutter dynamics and they made use of this property of sea clutter for the enhanced detection of a radar target embedded in sea clutter [Haykin and Li, 1995; Li, 1995; Li and Haykin, 1993, 1995]. Studies by Palmer et al. [1995] have also shown that sea clutter can be modeled as a low-dimensional attractor.

However, there has been a great deal of disagreement between the results on the chaotic characterization of sea clutter data reported by various researchers. For example, Leung and Haykin reported a D_2 value between 6 and 9 from their analysis using the GPA [Haykin and Leung, 1989, 1992; Leung and Haykin, 1990]. Haykin and Li reported a D_2 value between 7 and 9 [Haykin and Li, 1995]. Palmer et al. [1995] reported a D_2 value between 5 and 8. Li [1995] reported one positive Lyapunov exponent followed by a zero exponent and several negative exponents he also reported a Kaplan–Yorke dimension (D_{KY}) around 3, which is well below the estimated D_2 values. Ideally, D_{KY} should be closer to D_2 [Kaplan and Yorke, 1979; Kaplan et al., 1983; Ledrappier, 1981; Russell et al., 1980; Young, 1982, 1983]. These discrepancies are attributed to the types of algorithms used for estimating D_2 and Lyapunov exponents. The GPA for the estimation of D_2 and Wolf's algorithm for the estimation of the largest Lyapunov exponent require data with high signal-to-noise ratio (SNR), for the estimates to be reliable [Eckmann and Ruelle, 1992; Grassberger and Procaccia, 1983a & b; Wolf et al., 1985]. These algorithms are highly sensitive to the presence of noise in the collected data and produce spurious estimates of these parameters if the SNR is below 20 dB [Grassberger and Procaccia, 1983a & b; Wolf et al., 1985].

These discrepancies and the availability of other more "robust" algorithms for estimating D_2 and the Lyapunov spectrum prompted the senior author of this book to revisit the problem of chaotic characterization of sea clutter. In the experimental study reported in this book, the various tasks of interest have been solved using specific algorithms as indicated here:

1. Test of nonlinearity of the time series in three different ways:
   a. Using a wave-shaping filter, hereafter referred to as the WSF method
   b. Surrogate data analysis based on the growth of interpoint distances as the discriminating statistic [Schouten et al., 1994a], hereafter referred to as the SIPD method.
   c. Surrogate data analysis based on D_2 as the discriminating statistic [Theiler et al., 1992], hereafter referred to as the SCD method.
2. Correlation dimension estimation, using the maximum-likelihood principle as described by Schouten et al. [1994a], hereafter referred to as the STB_1 algorithm after its originators
3. Normalized embedding delay, using Shannon's mutual information criterion as described by Fraser and Swinney [Fraser, 1989; Fraser and Swinney, 1986], hereafter referred to as the MI algorithm
4. Global embedding dimension (d_E), using the method of false nearest neighbors described by Kennel et al, [1992], hereafter referred to as the GFNN algorithm
5. Local embedding dimension (d_L), using the method of local false nearest neighbors described in Abarbanel and Kennel [1993], hereafter referred to as the LFNN algorithm
6. Lyapunov spectrum, using the method described by Brown et al., hereafter referred to as the BBA algorithm after its originators [Briggs, 1990; Brown et al., 1991; Bryant et al., 1990]
7. Kolmogorov entropy estimation, using the maximum-likelihood principle as described by Schouten et al. [1994b], hereafter referred to as the STB_2 algorithm after its originators.

Detailed aspects of these algorithms are presented later.[1] The results presented in this book, obtained with these algorithms, demonstrate the

---

[1] The two specific pieces of software were used for the implementation of the algorithms: (a) RRCHAOS, developed at Delft University, The Netherlands, for the SIPD, STB_1, and STB_2 algorithms, and (b) cspX, developed by the Applied Chaos Group at Del Mar, CA, for the MI, GFNN, LFNN, and BBA algorithms.

chaotic dynamics of sea clutter in a very convincing way. The results were obtained using ground-truthed radar data collected from a series of extensive and thorough experiments carried out on the East Coast of Canada.

1.2 OBJECTIVES OF THE BOOK

In this book, we present a detailed account of the chaotic dynamics of sea clutter using real-life data, with the following aims in mind:

1. To provide systematic tests for the nonlinearity and stationarity of a chaotic time series
2. To describe preprocessing techniques for the preparation of real-life data for chaotic analysis without affecting the underlying dynamics of the process
3. To provide descriptions of the algorithms for reliable estimation of the following invariants of a strange attractor:
   a. Correlation dimension (D_2)
   b. Embedding delay (τ)
   c. Global embedding dimension (d_E)
   d. Local embedding dimension (d_L)
   e. Lyapunov spectrum (λ_i)
   f. Lyapunov (Kaplan–Yorke) dimension (D_{KY})
   g. Kolmogorov entropy (KE)
4. To demonstrate, in a collective manner, how the chaotic characterization of a physical (experimental) time series should be carried out.

The material presented herein appears in book form for the first time. Most importantly, it confirms the chaotic dynamics of sea clutter. Moreover, the book could be used as a model for how the chaotic dynamics of an experimental time series should be assessed.

LIST OF ABBREVIATIONS

A/D	Analog-to-digital (converter)
BBA	Brown, Bryant, and Abarbanel (algorithm)

BICE	Built-in calibration equipment
d_E	Embedding dimension
d_L	Local (dynamical) dimension
D_2	Correlation dimension
D_{KY}	Kaplan–Yorke dimension
D_{ML}	Maximum-likelihood estimate of correlation dimension
DFT	Discrete Fourier transform
DMA	Direct memory access
FFT	Fast Fourier transform
FIR	Finite-duration impulse response
FM	Frequency modulation
GPA	Grassberger–Procaccia algorithm
GFNN	Global false nearest neighbor
HH	Horizontal on both transmit and receive (polarizations)
HP	Horizon of predictability
IF	Intermediate frequency
IPIX	Intelligent pixel-processing
KE	Kolmogorov entropy
LFFN	Local false nearest neighbor
LPF	Low-pass filter
MI	Mutual information
PRF	Pulse repetition frequency
RBF	Radial basis function
RF	Radio frequency
SAW	Surface acoustic wave
SCD	Surrogate data analysis using correlation dimension as the discriminating statistic
SIPD	Surrogate data analysis using the interpoint distances as the discriminating statistic
SNR	Signal-to-noise ratio
STB	Schouten, Takens, and Bleek (algorithm)
STC	Sensitivity time control
TTL	Transistor–transistor–logic
VME	Versa module europa
VTS	Vessel traffic system
VV	Vertical on both transmit and receive (polarizations)
WGN	White Gaussion (random) noise
WSF	Wave-shaping filter

2

SEA CLUTTER

2.1 INTRODUCTION

Sea clutter, or *sea echo* refers to the backscattered returns from a patch of sea surface illuminated by a transmitted radar signal. The study of sea clutter is not only of theoretical but also of practical importance because it places severe limits on the detectability of radar returns from "point" targets on or near the surface. Such targets include low-flying aircraft, small marine vessels, navigation buoys, and pieces of ice floating in the ocean.

Our interest in this book is confined to the underlying dynamics of sea clutter acting on its own. More specifically, we are interested in sea clutter from a microwave radar operating at low grazing angles with the antenna dwelling in a fixed direction. In such a mode of operation, the *temporal* character of sea clutter is entirely due to the motion of the sea surface. This particular mode of operation is chosen in order to focus on the interaction between the incident electromagnetic wave and the sea surface solely as a function of time and in the simplest manner possible.[1]

The illuminated area of sea surface responsible for the radar backscatter (i.e., the *footprint* of the antenna) is determined by three factors:

[1] In a scanning mode of operation as in surveillance radar, the received signal assumes a more complicated form in that it is a function of both time and space.

1. Beamwidth of the antenna pattern
2. Height of the antenna above the sea surface
3. Nominal grazing angle of the incident electromagnetic wave (measured with respect to the sea surface), which is usually defined at the beam center (boresight).

Elevation

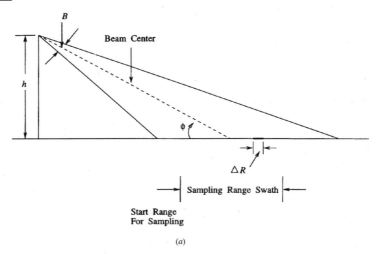

Plan

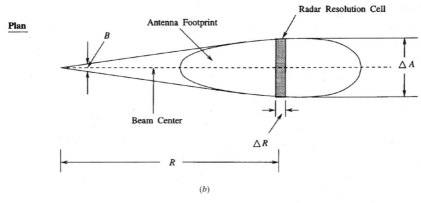

Figure 2.1 (*a*) Elevation and (*b*) plan views of antenna footprint illuminated by radar and the sampling area from which data are collected. Grazing angle ϕ and antenna beamwidth B shown are larger than actual values in radar experiments; this has been done here merely for the purpose of illustration, (h, antenna height; R, range; ΔR, range resolution; ΔA, cross-range resolution).

The antenna footprint may be subdivided in range into *resolution cells*, as illustrated in Fig. 2.1 for a pencil (cone-shaped) beam. The cross-range resolution of each cell is determined by the beamwidth of the antenna. The range resolution, denoted by ΔR, is defined as the minimum distance in range by which two point targets must be separated in order to be individually resolved. It is determined by the duration of the transmitted radar pulse, t_0, as shown by

$$\Delta R = \tfrac{1}{2} c t_0 \tag{2.1}$$

where c is the velocity of propagation. In the case of a radar using pulse compression, t_0 is the "effective" pulse duration. Thus, for a particular range (i.e., distance from the radar), the backscattered return is integrated over the area of the pertinent resolution cell. Note that the resolution cells are of unequal sizes in azimuth due to the spreading (with range) of the antenna footprint, as illustrated in Fig. 2.1.

In addition to the above-mentioned factors, the characterization of sea clutter depends on the following:

- *Air–sea interactions*, which are themselves influenced by the direction and speed of prevalent winds and also by the wind field at distant locations that generate waves that eventually propagate to the local area
- *Radar parameters*, namely, transmitted pulse duration, wavelength, and polarization on transmit and receive

2.2   SEA CLUTTER AT LOW GRAZING ANGLES

Basically, two types of surface waves influence radar backscatter at low grazing angles—capillary waves and gravity waves [Wetzel, 1970]:

- **Capillary waves** provide the fine structure of the sea surface. Their wavelengths are on the order of 2 cm or less. The dominant restoring force for capillary waves is surface tension.
- **Gravity waves** provide the larger and more visible structure of the sea surface. Their wavelengths vary from 200 m down to a small fraction of a meter. The dominant restoring force for gravity waves is the force of gravity, hence their name. Gravity waves may reside in one of two states: sea and swell. *Sea* is the state of waves when they

are blown by the wind that gave rise to them. *Swell* is the state of waves when they are no longer under the influence of the wind that generated them. Since the surface over which the waves travel acts as a low-pass filter, swell waves often take the form of long-crested low-frequency sinusoids.

In a microwave radar, the wavelength is of the order of a few centimeters or less. For example, the wavelength of an X-band radar (the type of the radar used in the study reported in this book), the wavelength is about 3.0 cm. For a simplified model of sea clutter produced by a microwave radar operating at low grazing angles, we may thus consider resonant scatter, or *Bragg scatter*, from the large number of small capillary waves riding on top of the gravity waves. What is being described here is a two-scale backscatter model [Apel, 1987; Wright, 1968] in which the surface waves act as a kind of diffraction grating, as illustrated in Fig. 2.2. From the grating equation of optics, the condition for *constructive* interference in the backscatter direction is defined by

$$n(\tfrac{1}{2}\lambda) = d \sin \theta_i \qquad (2.2)$$

where n is an integer, λ is the radar wavelength, d is the surface wavelength, and θ_i is the incidence angle (measured with respect to the normal to the sea surface); that is, θ_i is the complement of the grazing angle. For a low grazing angle, θ_i is close to $90°$ and $\sin \theta_i \approx 1$. Thus, the capillary

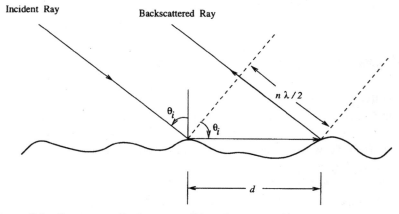

Figure 2.2 Bragg scattering condition from rough sea surface (θ_i, incidence angle; n, grating order; d, surface wavelength; λ, radar wavelength).

waves causing first-order ($n = 1$) Bragg scattering are those waves whose wavelengths are about one-half the radar wavelength.

The simplified Bragg scatter model described above ignores shadowing and diffraction effects that arise at low grazing angles. These effects make the relationship between the scattered field and the wave height profile of the sea surface highly nonlinear. The net result is that sea clutter is a nonlinear physical process.

2.3 COMPOUND *K*-DISTRIBUTIONS

Traditionally, sea clutter has been modeled as a stochastic (random) process, with the observable received signal from a radar resolution cell representing a sample function of that process. This approach was motivated by the classic paper of Rice [1945] on Gaussian noise theory. Longuet-Higgins [1962] generalized Rice's results to spatial as well as temporal variations in the modeling of rough surfaces.

When the pulse duration of a microwave radar is sufficiently short, the amplitude distribution of sea clutter has an extended tail, compared to the Rayleigh distribution that follows from Gaussian noise theory. This deviation was first recognized by Goldstein [1951]. In an effort to provide better fits to the amplitude distribution of sea clutter, ad hoc distributions, namely, lognormal [Trunk, 1972] and Weibull [Fay et al., 1977], were proposed during the 1970s. About the same time, non-Gaussian light experiments performed on layers of liquid crystals in turbulent motion were found to generate signals with statistical properties remarkably similar to those of sea clutter. It was then realized that the two-dimensional random-walk model [Beckmann and Spizzichino, 1963] being developed to explain those non-Gaussian light experiments could also form the statistical basis for addressing the sea clutter problem. Indeed, Jakeman and Pusey [1976] were the first to propose *K*-distributions, based on a modified Bessel function of the second kind, as a statistical model for sea clutter data. Among the various non-Gaussian models reported in the literature on the statistical characterization of scattered waves from rough surfaces, the *K*-distribution model stands out for two reasons: (i) phenomenological arguments and (ii) exact solutions to specific scattering problems. Jakeman and Tough [1988] developed a highly detailed statistical mechanical formulation for non-Gaussian compound Markov processes, with K-distributed noise treated as a special case.

In a physical sense, randomness is introduced into the K-distribution model for a continuous-medium scattering problem by invoking the following phenomenological arguments:

- The problem is formulated as one of scattering by a subfractal random-phase changing screen, which is defined as a diffusing layer that introduces spatially random variables in phase into the incident plane wave [Booker et al., 1950].

- As the distorted wavefront propagates beyond the diffusing layer, amplitude fluctuations are produced with characteristics governed by the statistical and correlation properties of that layer [Mercier, 1962; Salpeter, 1967].

An intuitively appealing way of looking at the K-distribution is as a Rayleigh process with its scale parameter varying as a gamma-distributed variate, which may reflect the Bragg scatter being modulated by the large-scale gravity waves. Another appealing feature of K-distributions is the fairly close fit they provide for the amplitude fluctuations of real-life sea clutter data over a wide range of experimental conditions [Nohara and Haykin, 1991]. However, for very high resolution radars the K-distribution model fails, hence the current research efforts to use higher compound models or spherically invariant processes [Drosopoulos, 1997].

In any event, the major limitation of the K-distribution model or extensions thereof is the fact that none of these models say anything about the underlying dynamics responsible for the generation of sea clutter. Information about the dynamics of sea clutter is particularly important in radar signal processing. For example, by using a nonlinear predictive model designed to capture the underlying dynamics of sea clutter, the target signal-to-clutter ratio at the output of the predictive model is made much larger than the corresponding value at the model input, and detectability of the target in sea clutter is thereby significantly improved [Haykin and Li, 1995]. The chaotic dynamics of sea clutter is the central theme of this book.

3

RADAR AND DATABASE

3.1 INTRODUCTION

In this chapter, we describe the radar and the field experiment details used in the experimental study reported in this book.

Radar is an electromagnetic system for the detection and location of objects. It operates by transmitting a particular type of waveform (e.g., a pulse-modulated sine wave) and analyzing the nature of the received signal. Radar is used to extend the capability of our own senses for observing the environment, especially the sense of vision. Important advantages of radar include the ability to do the following:

- Remotely detect targets of interest
- Track the movements of targets
- Estimate target parameters such as range and velocity

Moreover, the signal-processing operations are performed under varying and difficult environmental conditions: in all weather and during the day and night.

For a detailed treatment, the reader is referred to Skolnik [1980 and references therein].

13

3.2 IPIX RADAR

A large part of the database used for the study was collected using an instrument-quality radar called the intelligent pixel-processing (IPIX) radar [Haykin et al., 1991]. For more detailed information on the IPIX radar, interested readers may refer to [Hamburger, 1989; Krasnor, 1988, 1989]. Figure 3.1 presents a photograph of the IPIX radar. Since the time at which the experiments described herein were performed, the IPIX radar design has been modified. This section describes the system as it was at the time of the experiments used to collect the data used in this book.

The IPIX radar has the following system features and capabilities:

1. Dual polarized reception, pulse-to-pulse transmit polarization switching
2. Coherent transmission/reception
3. Pulse compression
4. X-band transmission
5. Digital data acquisition
6. Built-in calibration

Figure 3.1 Photograph of IPIX radar.

7. Flexible operation and modification

8. Computer control

3.2.1 System Features and Capabilities

Dual Polarization

The IPIX radar has two identical receivers; one is connected to the vertically polarized (V) antenna feed and the other to the horizontally polarized (H) antenna feed. Therefore, both polarizations are simultaneously received. A high-speed ferrite waveguide switch having 50 dB of isolation between channels is used to route the transmitted signal through either the horizontal or the vertical channel. To achieve this kind of isolation, a combination of three ferrite devices are actually used in the switch. The switch can change state at a continuous rate up to 2 kHz. Therefore, the radar is capable of near-simultaneous transmission of both orthogonal polarizations, which is necessary for measuring the full polarization scattering matrix of a target.

Coherent Transmission/Reception

The IPIX radar is a coherent radar allowing for accurate measurements of the phase of the returned radar echo. A high-stability 5-MHz crystal oscillator is used as the master reference clock for the entire system. Both intermediate-frequency (IF, 150 MHz) and radio frequency (RF, 9.24 GHz) sources are phase-locked to a master clock and are used to generate transmitted signal as well as to downconvert and demodulate the received signals. Quadrature demodulators in each receiver channel provide the in-phase (I-phase) and quadrature-phase (Q-phase) video outputs for subsequent digitization and processing.

Pulse Compression

The IPIX radar has a pulse compression subsystem consisting of a surface acoustic wave (SAW) pulse expander and matched SAW compressors in each of the two receivers. The expander generates a nonlinear frequency-modulated (FM) coded pulse 5 μs wide with a bandwidth of about 50 MHz. The compressed pulse has an effective width of 32 ns. The importance of pulse compression in IPIX is twofold. First, it makes IPIX a high-resolution radar with a range resolution of 4.8 m, allowing for improved visibility of small targets in clutter. Second, higher average power is delivered to the targets, increasing the maximum range at which a given target can be detected. Since we have no control of the location of targets of opportunity with respect to the radar, this increase in usable range is significant.

X-Band Transmission

The IPIX radar is an X-band radar that transmits a peak power of 8 kW at 9.39 GHz. The wavelength of the transmitted signal is approximately 3 cm. This wavelength couples well with the small capillary waves on the ocean surface, allowing for useful sea clutter measurements. Furthermore, at this wavelength, a target velocity of 1 knot causes a Doppler shift of about 34 Hz. At our maximum pulse repetition frequency (PRF) of 2 kHz, this allows for unaliased velocity measurements up to 30 knots.

Digital Data Acquisition

The data acquisition system is probably the most important aspect governing the utility of a radar for research work. Much effort was put into the design of this system in the IPIX radar. Digital acquisition in radar is challenging because of the high data sampling rates and the volume of data involved. Now, with the improved technology in data handling, the storage of the data is much less a problem.

The IPIX data acquisition system operates as follows: After each pulse is transmitted, the selected range interval is digitized at a rate of 30 MHz into the sweep buffer. Each range sample consists of the four receiver channels: horizontally and vertically polarized I and Q channels. Before the next pulse is transmitted, the sweep buffer is written through a high-speed Versa Module Europa (VME) bus Direct Memory Access (DMA) channel directly into the memory space of a 68020-based computer system. Each sweep is stored sequentially in memory. A significant benefit of this approach is that the data are available for processing immediately and are addressable by a simple pointer to memory. A high-capacity digital tape drive is used to store large numbers of data sets inexpensively either in raw form or after processing.

Built-in Calibration

The IPIX radar has built-in calibration equipment (BICE) consisting of a computer-controlled frequency-agile IF source that can be routed to the transmitter's up converter to produce an RF signal. This source, along with a digitally controlled attenuator, can then be used to inject known signals into the receiver in order to determine an input/output characteristic.

Sensitivity time control (STC) is also available to ensure that the received signal falls within the linear range of the receiver.

Flexible Operation and Modification

The IPIX radar has been designed so that it is flexible and modifiable. For flexible operation, remote-controlled coaxial switches are used throughout the system to allow computer control of the RF signal path. These switches make it a simple matter to configure the system for normal operation and for self-calibration and testing. Furthermore, hardware circuits monitor the state of these switches in order to ensure that they are in a safe configuration. The system is also easily modifiable. This is due to its modular design. All modules are rack mounted and easily accessible to allow for in-the-field modifications and repairs.

Computer Control

Virtually all aspects of the IPIX radar are changeable by computer control. These include:

- The mode of operation, which is effected by coaxial switches
- The transmit signal characteristics, which include pulse width, PRF, and polarization
- The sampling system, which specifies the azimuth and range windows required and the number of sweeps to store
- The antenna positioner unit, which controls the position and scan rate of the antenna
- The data storage facility and the generation of log files for each experiment

The computer interface for most of the control operations is a simple parallel bus that uses transistor–transistor–logic (TTL) signals and can be implemented with four 8-bit parallel ports. The control software is written in the C programming language for portability.

3.3   FIELD EXPERIMENTS

The radar was mounted in a fixed position on land 25–30 m above sea level. This height is typical of a ship-mounted radar. It was operated in a dwelling mode, with the antenna pointing along a fixed direction, illuminating a patch of the ocean surface.

Sea clutter data were recorded on different resolution cells defined within the antenna footprint (see Fig. 2.1). The start range values for sampling within the antenna footprint were 1200 m or greater for different experiments, and the range sampling rate varied from 5 to 30 MHz using an 8-bit analog-to-digital (A/D) converter for each of the four receiver channels:

- *Experiments at Cape Bonavista, Newfoundland*: The IPIX radar system was first installed at Cape Bonavista, Newfoundland, Canada, in May–June 1989. A wide variety of sea states can be observed at that site throughout the year. Sea clutter data for various sea states were collected. For each data set, I-channel (in-phase) and Q-channel (quadrature-phase) recordings were made, each consisting of a sampled range swath for at least 50,000 consecutive sweeps.

- *Experiments at Dartmouth, Nova Scotia*: Further field experiments at Dartmouth, Nova Scotia, Canada, were carried out in November 1993 using a refined version of the IPIX radar. Sea clutter (I and Q) data for various sea states, radar pulse durations, PRFs, different like polarizations were recorded.

- *Experiments at Argentia, Newfoundland*: For these experiments, a noncoherent commercial marine radar system was installed at a Vessel Traffic System (VTS) site in Argentia, Newfoundland, in January–February 1994. The radar data collected at this site contained only amplitude information, with fixed radar parameters.

Five sea clutter data sets collected from these radar experiments at different times of the day, at different months of the year, and under different environmental conditions were analyzed using the tools described later. The radar parameters and environmental conditions for the data sets were as follows:

1. *Data Set I (Dartmouth)*

Date	November 17, 1993
Time (local)	7:57 AM
Radar bearing angle	30°

Polarization	HH
Pulse repetition frequency	2000 Hz
Pulse duration	2000 ns
Beamwidth	1°
Range resolution	300 m
Area of first resolution cell	≈ 300 × 21 m
Range of sampling rate	5 MHz
Start range	1.2 km
Significant wave height	0.85 m
Wind	None

2. *Data Set II (Dartmouth)*

Date	November 14, 1993
Time (local)	3:13 PM
Radar bearing angle	105°
Polarization	VV
Pulse repetition frequency	2000 Hz
Pulse duration	200 ns
Beamwidth	1°
Range resolution	300 m
Area of first resolution cell	≈ 30 × 26 m
Range of sampling rate	10 MHz
Start range	1.5 km
Significant wave height	1.3 m
Wind	19 km/h from 90°

3. *Data Set III (Dartmouth)*

Date	November 17, 1993
Time (local)	9:38 PM
Radar bearing angle	190°
Polarization	VV
Pulse repetition frequency	1000 Hz
Pulse duration	200 ns
Beamwidth	1°
Range resolution	30 m
Area of first resolution cell	≈ 30 × 45 m
Range of sampling rate	10 MHz

Start range	2.6 km
Significant wave height	2.6 m
Wind	26 km/h from 320°

4. *Data Set IV (Cape Bonavista)*

Date	June 12, 1989
Time (local)	1:33 PM
Radar bearing angle	30°
Polarization	HH
Pulse repetition frequency	2000 Hz
Pulse duration	200 ns
Beamwidth	1°
Range resolution	30 m
Area of first resolution cell	$\approx 30 \times 99$ m
Range of sampling rate	30 MHz
Start range	5.7 km
Significant wave height	2.5 m
Wind	31 km/h from 120°

5. *Data Set V (Argentia)*

Date	February 8, 1994
Time (local)	4:26 AM
Radar bearing angle	270°
Polarization	HH
Pulse repetition frequency	1300 Hz
Pulse duration	250 ns
Beamwidth	1.3°
Range resolution	37.5 m
Area of first resolution cell	$\approx 37.5 \times 58$ m
Range of sampling rate	8 MHz
Start range	3.3 km
Significant wave height	2.0 m
Wind	36 km/h from 190°

The time histories of amplitude versus range for these five data sets are displayed in Figs. 3.2a–e, respectively.

Figure 3.2 shows sample time histories of the received signal amplitude versus range for five different situations. Figure 3.2a shows the amplitude plot for a radar bearing angle of 30° and horizontal polarization on both transmit and receive (hereafter referred to as horizontal like-polarization, or HH). There was no wind, and the wave height was 0.85 m. As expected, there is no particular pattern seen in this first amplitude plot. Figure 3.2*b* shows the amplitude plot for a radar bearing angle of 105° and vertical polarization on both transmit and receive (hereafter referred to as vertical like-polarization, or VV). The winds were 19 km/h from 90°, and the wave height was about 1.3 m. The swell waves are clearly visible in this

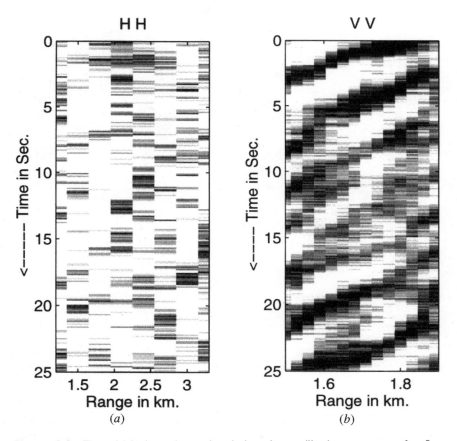

Figure 3.2 Time histories of received signal amplitude vs. range for five different situations: (*a*) $\Delta R = 300.0$ m, range sampling rate 5 MHz, wave height 0.85 m; (*b*) $\Delta R = 30.0$ m, range sampling rate 10 MHz, wave height 1.3 m.

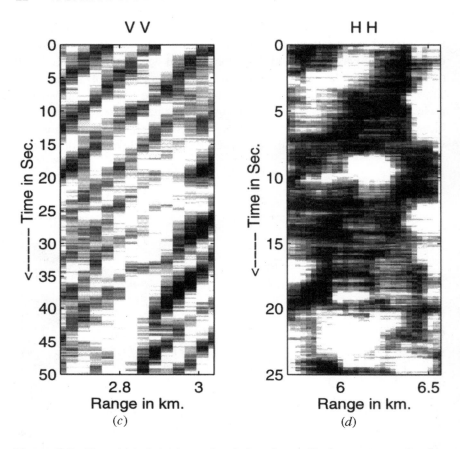

Figure 3.2 Time histories of received signal amplitude vs. range for five different situations: (*c*) $\Delta R = 30.0$ m, range sampling rate 10 MHz, wave height 2.6 m; (*d*) $\Delta R = 30.0$ m, range sampling rate 30 MHz, wave height 2.5 m;

second figure: an expected result as the radar was looking closely into the swell and the wind direction. Figure 3.2*c* shows the amplitude plot for a radar bearing angle of 190° and vertical like-polarization. The winds were 26 km/h from 320°. This situation shows some indication of the swell waves as the radar is looking essentially downwind. Figure 3.2*d* shows the amplitude plot for a radar bearing angle of 30° and horizontal like-polarization. The winds were 31 km/h from 120°, and the wave height was 2.5 m. This fourth figure shows no particular wave pattern: again an

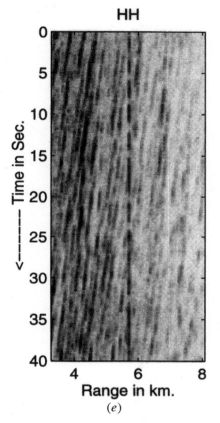

HH

Figure 3.2 Time histories of received signal amplitude vs. range for five different situations: (e) $\Delta R = 37.5$ m, range sampling rate 8 MHz, wave height 2.0 m;

expected result as the radar was looking essentially cross-wind and cross-swell. Finally, Fig. 3.2e shows the amplitude plot for a radar bearing angle of 270° and horizontal like-polarization. The winds were 36 km/h from 190°, and the wave height was 2.0 m. The swell waves are clearly visible in Fig. 3.2e. Note, however, this latter figure is on a much larger range scale than that of Fig. 3.2b, hence the difference in their appearance. The continuous response at the center range is from a moored buoy.

4

CHAOTIC DYNAMICS OF A PHYSICAL TIME SERIES

4.1 INTRODUCTION

Nonlinear dynamical systems can exhibit behavior that has a disordered appearance despite the deterministic nature of the underlying physics. Such chaotic behavior, which is now known to be common, can arise under proper conditions in most nonlinear systems.

The dynamics of physical systems can be represented concisely by geometric objects in a multidimensional physical phase space or state space, for example, trajectories, attractors, basin of attraction, and unstable manifolds. Much progress has been made in mathematically describing chaotic dynamics by connecting experimental time series to those geometric objects. This technique consists of forming a mapping between the observed time series and vectors in some simple multidimensional space. These vectors then form points on a trajectory and the multi-dimensional space becomes a representation of the actual physical phase space of the system. This process is termed *embedding*.[1] The main theorems connecting these two realms of measurement and physical phase

[1] A detailed description of embedding is presented in a later chater of this book.

space are Takens's theorem [Takens, 1981], its precursors [Mañé, 1981], and its sequels [Sauer et al., 1991].

More recently, the study of chaotic dynamics has entered a new phase. In addition to the original pursuits of demonstrating chaos in a wide range of situations and studying the properties of chaotic dynamics, many researchers are interested in the general set of problems that involve chaos. By the term "chaos" we mean utilizing the basic knowledge of the theory of chaos to analyze experimental time series data. In most of these cases, the equations underlying the dynamics are either not known or are known in principle but are too complex to solve. In most situations we are given observations of only one or two variables of a system under investigation, and the requirement is to confirm the chaotic nature of the system.

Many of the current methods in nonlinear signal processing have been conceived due to the entirely new idea of deterministic chaos. These innovative techniques have provided experimentalists with new ways of understanding the implications of their data.

It is an important concept in dynamics that dissipative dynamical systems are typically characterized by the presence of attracting sets or attractors [Abarbanel, 1996; Abarbanel et al., 1993; Baker and Gollub, 1996; Lorenz, 1963; Mandelbrot, 1983; Moon, 1992; Ott, 1993; Schuster, 1988]. Thus, the trajectories of a dynamical system converge toward a limit set, an attractor, in the state space, which only fills a subset of the space. Because of the assumptions of boundedness and nonperiodicity, the attractor necessarily has a very peculiar and intricate structure with possibly fractal properties, a situation referred to as strange attractor (footprints of chaos) [Abarbanel, 1996; Abarbanel et al., 1993; Baker and Gollub, 1996; Lorenz, 1963; Mandelbrot, 1983; Moon, 1992; Ott, 1993; Peitgen et al., 1992]. The existence of a fractal dimension for an attractor can therefore be used as an indication of possibly chaotic dynamics of the underlying physical system, with the dimension defining a lower bound on the number of degrees of freedom required to characterize the system. Another consequence of the lack of periodicity is the broadband spectrum of a chaotic signal, which is a necessary condition for the assessment of chaos. Thus, the chaotic characterization of a dynamical system can be achieved by studying the attractor of the system in a multidimensional state space [Mandelbrot, 1977].

There are different fractal dimensions associated with strange attractors. They are the capacity dimension or box-counting dimension (D_0), information dimension (D_1), correlation dimension (D_2), and Lyapunov dimension (D_{KY}). Of all these dimensions, we pick up only the correlation dimension and Lyapunov dimension in our studies. Detailed descriptions of the algorithms used for the estimation of these two quantities are given in the following sections of this chapter.

The Lyapunov exponents associated with a trajectory are essentially a measure of the average rates of expansion and contraction of trajectories surrounding it. They are asymptotic quantities, defined locally in phase space, and describe the exponential rate at which a perturbation to a trajectory of a system grows or decays with time at a certain location in the phase space. We describe in detail the methods adopted in the literature for estimation of Lyapunov exponents of chaotic systems.

Yet another important characteristic of the chaotic dynamics is its Kolmogorov entropy. It is a measure of the rate of information loss along the attractor or a measure of the degree of predictability of points along the attractor given any arbitrary initial point on the attractor. A positive and finite entropy is considered to be the conclusive proof of a system to be chaotic. A regular and cyclic process is characterized by its zero entropy and a stochastic or random process is characterized by an infinite entropy.

Estimation from time series of descriptive measures such as dimensions, Lyapunov exponents, or Kolmogorov entropy derived from the theory of deterministic chaos is well established in the case of data generated by low-dimensional deterministic dynamical systems in numerical and laboratory experiments.

The first significant investigation into the behavior of simple chaotic systems was performed by Lorenz in 1963. He studied numerically the behavior of a simplified model describing convective heat flow and observed chaotic self-excited oscillations. After his paper was published, physicists as well as mathematicians became increasingly interested in chaotic dynamics. Later theoretical studies proved that chaos occurs as a feature of orbits arising from nonlinear evolution rules, which are systems of differential equations with three or more degrees of freedom or invertible discrete time maps with two or more degrees of freedom [Guckenheimer and Holmes, 1983; Moon, 1992; Thomson and Stewart, 1986]. The reason for this requirement is beyond the scope of this book, but it turns out that with the parameters in the equations set such that the system becomes chaotic, it is mathematically impossible to solve the differential equations to get an expression for the long-term behavior of the system. The nonlinearity of the observed time series is an essential requirement for the underlying system to be chaotic.

Chaotic dynamics and fractal geometry have a close relationship in that one of the hallmarks of chaotic behavior is the manifestation of fractal geometry, particularly for strange attractors in dissipative systems. For a practical definition, we take a strange attractor to be an attracting set with fractal dimension. This means that the geometric structures generated by chaotic systems are extremely complex and fractal. Regions in phase space are stretched, contracted, folded, and remapped into a compact region of the original space whose volume shrinks to zero for a dissipative system, leaving gaps in the phase space. The orbits tend to fill up less than an integer subspace in phase space.

The simplest way to understand the difference between classical dynamics and chaotic dynamics is to imagine two systems starting from the same initial state but with some infinitesimal difference between them. In a classical system, one would expect that the infinitesimal difference would lead to small difference in the states of the system at a future time, with the difference tending to grow linearly with time. In contrast, in a chaotic system, the discrepancy between the two trajectories grows exponentially. Thus, for short times the systems will remain in close agreement, but soon thereafter they will diverge radically and it will become impossible to observe any correlation between the two. This property is known as the sensitive dependence on initial conditions. Another way to think about this is to say that no matter how well the current state of the system is known, it is impossible to tell which path it will take even a moderate amount of time into the future.

4.2 SOME BASIC DEFINITIONS

4.2.1 Attractors

In nature, all processes seem to tend toward some sort of stable state or equilibrium. This state is called an attractor in mathematical terms. Attractors are geometric forms that characterize the long-term behavior of a dynamical process in the state space. An attractor is a set of states (points in the phase space), invariant under the dynamics, toward which neighboring states in a given basin of attraction asymptotically approach in the course of dynamic evolution. An *attractor* is defined as the smallest unit that cannot be decomposed into two or more attractors with distinct basins of attraction. This restriction is necessary since a dynamical system may have multiple attractors, each with its own basin of attraction.

Conservative systems do not have attractors, since the motion is periodic. In dissipative dynamical systems, however, volumes shrink exponentially and hence attractors have zero volume in n-dimensional phase space. A stable fixed point surrounded by a dissipative region is an attractor known as a sink. Regular attractors (corresponding to zero Lyapunov exponents) act as limit cycles, in which trajectories circle around a limiting trajectory that they asymptotically approach but never reach. These simple attractors are presented in Fig. 4.1.

For years, scientists have believed that the attractors toward which physical systems tend are quite simple. They should be either rest points or equilibrium points such as the rest position of a pendulum, or else they should be limit cycles or repeating configurations, such as often occur in population biology or economic systems. But this view changed after the discovery of strange attractors during the last few decades.

The attractor associated with a chaotic motion in phase space is not a simple geometrical object such as a finite number of points, a closed curve, or a torus. In fact, it is not even a smooth surface; that is, it is not a manifold. They are called "strange attractors" and are complicated geometrical objects that possess fractal dimensions. In the late 1960s, Stephen Smale [1967] proposed several geometric models of new attractors that were much more complicated than fixed points or limit cycles. The dynamics of these attractors was chaotic, not steady state or periodic. Later R. F. Williams [1970] showed that these attractors were stable in the sense that small changes in the basic system could not destroy them.

So, strange attractors exist as mathematical objects, but do they exist in nature? E. N. Lorenz, a meteorologist at Massachussets Institute of Technology (MIT), found an answer to this question in the early sixties, with a simplified model to metereological turbulence, which is chaotic. His model was a system of differential equations in three dimensions. When viewed on a computer graphics terminal, each solution curve of the system tends toward the same object—the Lorenz attractor. A Lorenz attractor in three-dimensional space is shown in Fig. 4.2. Once near the attractor, the solution oscillates about one of two lobes of the attractor. The important point is that the number of oscillations about each lobe seems to be random and depends very much on which solution curve is followed. This, of course, is chaos.

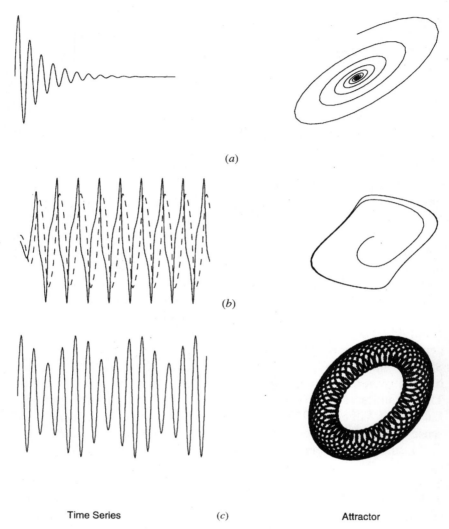

Time Series   *(c)*   Attractor

Figure 4.1 Some simple attractors: *(a)* point attractor; *(b)* limit cycle; *(c)* 2-torus.

Consequently, the Lorenz system never settles down to equilibrium or periodic behavior but, rather, continuously cycles chaotically. Since the Lorenz discovery, similar strange attractors have been found in a variety of scientific disciplines, including the Hénon attractor and the Rossler attractor [Hénon, 1976; Roessler, 1976].

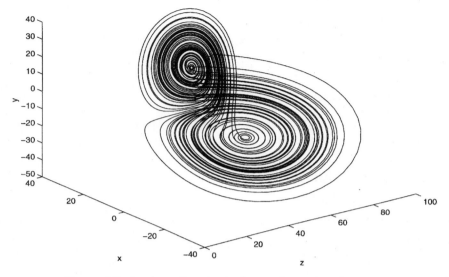

Figure 4.2   Lorenz attractor in three-dimensional space.

While all chaotic attractors are strange, not all strange attractors are chaotic [Grebogi et al., 1981]. Strange and nonchaotic attractors have been found in quasiperiodically forced systems [Ding et al., 1989; Grebogi et al., 1984; Romeiras et al., 1987; Romeiras et al., 1989]. In these cases, the attractors have a fractal structure but are not associated with any positive Lyapunov exponents (presence of at least one positive Lyapunov exponent is a necessary requirement for a system to be chaotic).

In contrast with the spectra of periodic and quasiperiodic attractors, which consist of a number of sharp spikes, the spectrum of a chaotic signal has continuous broadband characteristics. In addition to the broadband component, the spectrum of a chaotic signal often contains spikes that indicate the predominant frequencies in the signal.

4.2.2   Basin of Attraction

The *basin of attraction* is the set of points in the space of the system variables such that initial conditions chosen in this set dynamically evolve to a particular attractor.

4.2.3 Sink (Map)

A *sink* is defined as a stable point of a map which, in a dissipative dynamical system, is an attractor.

4.2.4 Limit Cycle

A *limit cycle* is an attracting set to which orbits or trajectories converge and upon which trajectories are periodic.

4.2.5 Strange Attractor

An attracting set that has zero measure in the embedding state space and has fractal dimension is termed a *strange attractor*. Trajectories within a strange attractor appear to skip around randomly.

Strange attractors arise from nonlinear dynamical systems. Physically, a dynamical system is nothing but a system that moves. Mathematically, a dynamical system is defined by a state space that describes the instantaneous states available to the system and an evolution operator that tells how the state of the system changes in time. This evolution operator is sometimes termed the physics of the system.

4.2.6 Manifold

Rigorously, an *n*-dimensional (topological) *manifold* is a topological space M such that any point in M has a neighborhood $U \subset M$ that is homeomorphic to *n*-dimensional Euclidean space. The homeomorphism is called a chart, since it lays that part of the manifold out flat, like charts of regions of Earth. So, a preferable statement is that any object that can be "charted" is a manifold.

The most important manifolds are differentiable manifolds. These are manifolds where overlapping charts "relate smoothly" to each other, meaning that the inverse of one followed by the other is an infinitely differentiable map from Euclidean space to itself.

4.2.7 Fractal Dimension

The dimension of an attractor is clearly the first level of knowledge necessary to characterize its properties. Generally speaking, we may think of the dimension as giving, in some way, the amount of information necessary to specify the position of a point on the attractor to within a given accuracy. The dimension is also a lower bound on the number of essential variables needed to model the dynamics.

For simple attractors, defining and determining the dimension is easy. For example, using any reasonable definition of dimension, a stationary time-independent equilibrium (fixed point) has a dimension zero, a stable periodic oscillation (limit cycle) has dimension 1, and a doubly periodic attractor (2-torus) has dimension 2. It is because their structure is very regular that the dimension of these simple attractors takes on interger values.

Chaotic (strange) attractors, however, often have a structure that is not simple; they are often not manifolds and have a highly fractured character. For strange attractors, intuition based on properties of regular, smooth examples does not apply. The most useful notions of dimension take on values that are typically not integers. These noninteger dimensions are called *fractal dimensions*.

To fully understand the properties of a chaotic (strange) attractor, one must take into account not only the attractor itself but also the "distribution" or "density" of points on the attractor. This is because chaotic attractors typically have complex and irregular structure against regular and simple structures for attractors such as fixed point, limit cycle, and 2-torus. For any attractor, the dimension can be estimated by looking at the way in which the number of points within a sphere of radius r scales as the radius shrinks to zero. The geometric relevance of this observation is that the volume occupied by a sphere of radius r in dimension d behaves as r^d.

For regular attractors, irrespective of the origin of the sphere, the dimension would be the dimension of the attractor. But for a chaotic attractor, the dimension varies depending on the point at which the estimation is performed. (This is because of the highly irregular structure of the attractor.) To have some dimension that is invariant under the dynamics of the process, we will have to average the point densities of the attractor around it. For the purpose of identifying the dimension in this fashion, we find the number of points $\mathbf{y}(k)$ within a sphere around some phase space location $\mathbf{x}$. This is defined by

$$n(\mathbf{x}, r) = \frac{1}{N} \sum_{k=1}^{N} \theta(r - |\mathbf{y}(k) - \mathbf{x}|) \qquad (4.1)$$

where θ is the Heaviside function.

This counts all the points on the orbit $\mathbf{y}(k)$ within a radius r from the point $\mathbf{x}$ and normalizes that number by the total number of points N in the data. Also, we know that the point density [denoted by $\rho(\mathbf{x})$] on an attractor need not be uniform (for a strange attractor) on the figure of the attractor. In this kind of situation, as we know, it would be revealing to look at the moments of the function $n(\mathbf{x}, r)$. This reveals different aspects of the distribution of the points.

Choosing the function as $n(\mathbf{x}, r)^{q-1}$ and defining the function $C(q, r)$ of two variables q and r by the mean of $n(\mathbf{x}, r)^{q-1}$ over the attractor weighted with the natural density $\rho(\mathbf{x})$ yield

$$
\begin{aligned}
C(q, r) &= \int d^d x \, \rho(\mathbf{x}) n(\mathbf{x}, r)^{q-1} \\
&= \frac{1}{M} \sum_{k=1}^{M} \left[\frac{1}{K} \sum_{n=1, n \neq k}^{K} \theta(r - |\mathbf{y}(n) - \mathbf{y}(k)|) \right]^{q-1}
\end{aligned}
\tag{4.2}
$$

The quantity $C(q, r)$ is called the "correlation function" on the attractor. It is a measure of the probability that two points $\mathbf{y}(n)$ and $\mathbf{y}(k)$ on the attractor are separated by a distance r. The values M and K are large but not infinite. This function of two variables is an invariant on the attractor, but it has become conventional to look only at the variation of this quantity when r is small. In that limit, it is *assumed* that

$$
C(q, r) \approx r^{(q-1)D_q}
\tag{4.3}
$$

defining the generalized fractal dimension D_q when it exists. From the above equation, D_q can be estimated in the limiting case as

$$
D_q = \lim_{r \text{ small}} \frac{\log[C(q, r)]}{(q - 1) \log[r]}
\tag{4.4}
$$

In practice, we need to compute $C(q, r)$ for a range of small r over which we can argue that the function $\log[C(q, r)]$ is linear in $\log[r]$ and then pick off the slope over the range.

Box-Counting Dimension (D_0)

This is called the "*box-counting dimension*" because it is being estimated as the number of spheres of radius r, namely, how many boxes do we need to cover all the points in the data set? If we evaluate this number $N(r)$ as a function of r as becomes small, then

$$D_0 = \lim_{r \text{ small}} \frac{\log[N(r)]}{\log[1/r]} \qquad (4.5)$$

This can also be defined as

$$D_0 = \lim_{q \to 0} D_q \qquad (4.6)$$

Information Dimension (D₁)

The *information dimension* D_1 is a generalization of the capacity that takes into account the relative probability of cubes used to cover the set. This dimension was originally introduced by Balatoni and Renyi [1959]. From the generalized dimensions, this can be defined as

$$D_1 = \lim_{q \to 0} D_q \qquad (4.7)$$

The limit $q \to 1$ results in an indeterminate form of the above equation. We can use the well-known L'Hospital's rule to solve it.

Correlation Dimension

For $q = 2$, the definition of the fractal dimension D_q in Eq. (4.4) assumes a simple form that lends it to reliable computation. The resulting dimension, D_2 is called the *correlation dimension* of the attractor [Grassberger and Procaccia, 1983a,b] and is estimated as the slope of the log–log curve as follows:

$$D_2 = \lim_{r \text{ small}} \frac{\log[C(2, r)]}{\log[r]} \qquad (4.8)$$

The correlation dimension is interesting because it is relatively easy to determine from experimental data. The basic idea was introduced by Grassberger and Procaccia [1983a,b] and independently by Takens [1983a].

4.2.8   Lyapunov Spectrum

One of the defining attributes of an attractor for which the dynamics is chaotic is that it displays exponentially sensitive dependence on initial conditions [Benettin et al., 1980a,b; Eckmann and Ruelle, 1985; Farmer,

1985; Mayer-Kress and Hubler, 1986; Oseledec, 1968; Wolf et al., 1985].
An illustration of sensitive dependence on initial conditions is shown in
Fig. 4.3. Consider two nearby initial conditions $s_1(0) = s_0$ and
$s_2(0) = s_0 + \varepsilon(0)$ and imagine that they are evolved forward in time by a
continuous-time dynamical system yielding orbits $s_1(t)$ and $s_2(t)$. At time
t, the separation between the two orbits is $\varepsilon(t) = s_1(t) - s_2(t)$. Suppose that
for small $|\varepsilon(0)|$ and for large t we have

$$\frac{|\varepsilon(t)|}{|\varepsilon(0)|} \approx e^{\lambda t} \tag{4.9}$$

If the parameter λ, denoting the largest Lyapunov exponent, is positive,
then we say that the system displays sensitive dependence on initial
conditions and is chaotic [Benettin et al., 1980a,b; Eckmann and Ruelle,
1985; Farmer, 1985; Mayer-Kress and Hubler, 1986; Oseledec, 1968;
Wolf et al., 1985]. This property drastically limits the possibility of long-
term prediction. Sensitivity to initial conditions leads to the fact that
nearby trajectories diverge/converge exponentially with respect to time.
For an n-dimensional chaotic system, there are n Lyapunov exponents
(positive, zero, and negative) and a measure of the divergence/con-
vergence of its trajectories in the state space is provided by the
positive/negative Lyapunov exponents, respectively [Wolf et al., 1985].
The set of all Lyapunov exponents of the dynamical system is termed
the *Lyapunov spectrum*.

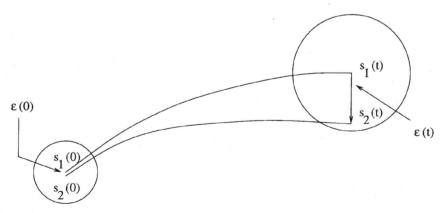

Figure 4.3   Illustration of sensitive dependence on initial conditions.

To understand the significance of the spectrum of Lyapunov exponents, consider the effects of the dynamics on a small spherical fiducial hyper-volume in the state space. Arbitrarily complicated dynamics, like those associated with chaotic systems, can cause the fiducial element to evolve into extremely complex shapes. However, for small enough length scales and short time scales the initial effect of the dynamics will be to distort the evolving sphere into an ellipsoidal shape, with some directions being stretched and others contracted in the multidimensional space. This is depicted in Fig. 4.4. The principal axis of the ellipsoid corresponds to the most unstable direction of the flow. The asymptotic rate of expansion of this axis is what is measured by the largest Lyapunov exponent. More precisely, if the infinitesimal radius of the initial fiducial volume is denoted by $r(0)$ and the length of the ith principal axis at time t is denoted by $l_i(t)$, then the ith Lyapunov exponent can be defined as

$$\lambda_i = \lim_{t \to \infty} \frac{1}{t} \log \frac{l_i(t)}{r(0)} \qquad (4.10)$$

By convention, the Lapunov exponents are always ordered so that $\lambda_1 > \lambda_2 > \lambda_3 > \cdots > \lambda_n$.

Usually, the Lyapunov exponents are measured in *nats per second*, where *nat* is the natural unit of information. In our present study, we present the Lyapunov exponents in *nats per sample*. This is nothing but

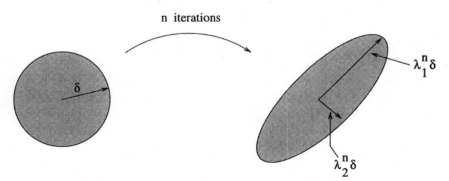

Figure 4.4 Effect of stretching and folding of dynamics of infinitesimal spherical volume in state space.

the exponents in *nats per second* normalized with respect to the PRF of the radar. That is,

$$\frac{\lambda}{\text{samples/s}} = \frac{\text{nats/s}}{\text{samples/s}} = \text{nats/sample} \tag{4.11}$$

How to Calculate the Lyapunov Spectrum

The spectrum of Lyapunov exponents is determined by following the evolution of small perturbations to an orbit by the linearized dynamics of the system whose dynamics is represented by the equation

$$\mathbf{y}(k+1) = \mathbf{F}(\mathbf{y}(k)) \tag{4.12}$$

Suppose we have established by state space reconstruction or direct observation an orbit $\mathbf{y}(k)$, $k = 1, 2, \ldots, N$, to which we make a small change/perturbation $[\mathbf{w}(1)]$ at "time" unity. The evolution of dynamics will now be

$$\mathbf{y}(k+1) + \mathbf{w}(k+1) = \mathbf{F}(\mathbf{y}(k) + \mathbf{w}(k)) \tag{4.13}$$

where $\mathbf{w}(k)$ can be estimated from the above equation in the following way if $\mathbf{w}(k)$ starts small and stays small. The following equation is true for these conditions:

$$\begin{aligned} \mathbf{w}(k+1) &= \mathbf{F}(\mathbf{y}(k)) - \mathbf{y}(k+1) + \mathbf{DF}(\mathbf{y}(k)) \cdot \mathbf{w}(k) + \cdots \\ &= \mathbf{DF}(\mathbf{y}(k)) \cdot \mathbf{w}(k) \end{aligned} \tag{4.14}$$

The evolution of the perturbation can be written as

$$\begin{aligned} \mathbf{w}(L+1) &= \mathbf{DF}(\mathbf{y}(L)) \cdot \mathbf{DF}(\mathbf{y}(L-1)) \cdots \mathbf{DF}(\mathbf{y}(1)) \cdot \mathbf{w}(1) \\ &= \mathbf{DF}^L(\mathbf{y}(1)) \cdot \mathbf{w}(1) \end{aligned} \tag{4.15}$$

where $\mathbf{DF}^L(\mathbf{y}(1))$ is the shorthand notation for the composition of L Jacobian matrices $\mathbf{DF}(\mathbf{x})$ along the orbit $\mathbf{y}(k)$.

In 1968, Oseledec proved the important multiplicative ergodic theorem, which includes a demonstration that the eigenvalues of the orthogonal matrix $\mathbf{DF}^L(\mathbf{x}) \cdot [\mathbf{DF}^L(\mathbf{x})]^T$ are such that the matrix

$$\lim_{L \to \infty} \{\mathbf{DF}^L(\mathbf{x}) \cdot [\mathbf{DF}^L(\mathbf{x})]^T\}^{(1/2)L} \tag{4.16}$$

exists and has eigenvalues $\exp[\lambda_1]$, $\exp[\lambda_2]$, ..., $\exp[\lambda_n]$ for an n-dimensional dynamical system and which are independent of $\mathbf{x}$ for almost all $\mathbf{x}$ within the basin of attraction of the attractor. The λ_i's are the global

Lyapunov exponents. Their independence of where one starts within the basin of attraction means that they are a characteristic of the dynamics, not the particular observed orbit.

Oseledec also proved the existence of the eigendirections of $\mathbf{DF}^L(\mathbf{y}(1))$. There are linear invariant manifolds of the dynamics: Points along these eigendirections stay along those directions under the action of the dynamics as long as the perturbation stays small. In 1979, Ruelle [1979] and others extended this idea to the nonlinear manifolds and to more complicated spaces. These eigendirections depend on where one is on the attractor, not on where one starts in order to get there. That is, if we want to know the eigendirections of $\mathbf{DF}^L$ at some point $\mathbf{y}(L)$ and we begin at $\mathbf{y}(1)$ and take L steps, or begin at $\mathbf{y}(2)$ and take $L - 1$ steps, the eigendirections will be the same as long as L is large enough. In practice, on the order of 10 steps ($L = 10$) are usually required to establish the eigendirections correctly [Abarbanel et al., 1993].

The multiplicative ergodic theorem of Oseledec is a statement about the eigenvalues of the dynamics in the tangent space [Eckmann and Ruelle, 1985] to $\mathbf{x} \rightarrow \mathbf{F}(\mathbf{x})$ and is motivated by a statement about linearized dynamics of the mapping. It is also a characterization via the λ_i of nonlinear properties of the system. Basically it is an analysis of the behavior of the nonlinear system in the neighborhood of an orbit on the strange attractor. By following the tangent space behavior (the linearized behavior near an orbit) locally around the attractor, we extract statements about the global nonlinear behavior of the system. A compact geometric object such as a strange attractor, which has positive Lyapunov exponents, cannot arise in a globally linear system. The stretching associated with the unstable directions of $\mathbf{DF}^L(\mathbf{y}(k))$ locally and the folding associated with the dissipation are the required ingredients of the theorem.

If any of the λ_i's are positive, then small perturbations will grow exponentially quickly over the attractor. Positive λ_i's are the hallmark of chaotic behaviour. If all λ_i's are negative, then a perturbation will decrease to zero exponentially quickly, and all the orbits are stable. In a dissipative system, the sum of all the λ_i's must be negative, and the sum governs the rate at which volumes in state space shrink to zero. If the underlying dynamics is governed by a differential equation, one of the λ_i's will be zero. This corresponds to a perturbation directly along the vector field, and such a perturbation simply moves one along the same orbit on which one started, so nothing happens in the long run. Indeed, one can tell, in principle, that if the source is governed by a coupled system of nonlinear differential equations, then it is characterized by the presence of a zero global Lyapunov exponent.

If one knows the vector field, determination of λ_i's is straightforward, though numerically it presents some challenges [Greene and Kim, 1987]. The main challenge comes from the fact that although each $\mathbf{DF}(\mathbf{y})$ has eigenvalues more or less like $\exp[\lambda_i]$, the composition of L Jacobians, $\mathbf{DF}^L(\mathbf{y})$, has eigenvalues of approximately $\exp[\lambda_1 L]$, $\exp[\lambda_2 L], \ldots$, $\exp[\lambda_n L]$, and since $\lambda_1 > \lambda_2 > \cdots > \lambda_n$, as L becomes large, the matrix is terribly ill conditioned. The diagonalization of such a matrix presents serious numerical problems. Standard $\mathbf{QR}$ decomposition routines do not work as well as one would like, but a recursive $\mathbf{QR}$ decomposition due to Eckmann et al. [1986] does the job. The problem in the direct $\mathbf{QR}$ decomposition is keeping track of the orientation of the matrices from one step to the next, so it is proposed to write each Jacobian in its $\mathbf{QR}$ decomposition as

$$\mathbf{DF}(\mathbf{y}(i)) \cdot \mathbf{Q}(i-1) = \mathbf{Q}(i) \cdot \mathbf{R}(i) \tag{4.17}$$

where $\mathbf{Q}(0) = \mathbf{I}$, an identity matrix. This would give the first part of the product:

$$\mathbf{DF}(\mathbf{y}(1)) = \mathbf{Q}(1) \cdot \mathbf{R}(1)$$
$$\mathbf{DF}(\mathbf{y}(2)) \cdot \mathbf{Q}(1) = \mathbf{Q}(2) \cdot \mathbf{R}(2)$$
$$\mathbf{DF}(\mathbf{y}(2)) = \mathbf{Q}(2) \cdot \mathbf{R}(2) \cdot \mathbf{Q}(1)^{\mathrm{T}} \tag{4.18}$$
$$\mathbf{DF}(\mathbf{y}(2)) \cdot \mathbf{DF}(\mathbf{y}(1)) = \mathbf{Q}(2) \cdot \mathbf{R}(2) \cdot \mathbf{R}(1)$$

Continuing, we have, for $\mathbf{DF}^L(\mathbf{y}(1))$, the following expression:

$$\mathbf{DF}^L(\mathbf{y}(1)) = \mathbf{Q}(L) \cdot \mathbf{R}(L) \cdot \mathbf{R}(L-1) \cdots \mathbf{R}(1) \tag{4.19}$$

This expression is easily diagonalized, as the product of upper right triangular matrices is an upper triangular matrix and the eigenvalues of such a matrix are the numbers along the diagonal. The Lyapunov exponents are read off the product of the upper triangular matrices.

If we have differential equations rather than mappings, or more precisely, if we want to let the step in the map approach zero, we can simultaneously solve the differential equation and the variational equation for the perturbation around the computed orbit. Then a recursive $\mathbf{QR}$ procedure can be used for determining the Lyapunov exponents.

The problem of determining the Lyapunov exponents when only scalar observations $s(n)$, $n = 1, 2, \ldots, N$, are made is another challenge. As always the first step is reconstruction of the phase space by the embedding technique. Once we have our collection of embedded vectors, there are two general approaches to estimate the exponents:

1. An analytic approach when a predictive model of the underlying process is available for supplying values for the Jacobians of the dynamical rules.
2. In the case where we do not have such information about the dynamics, we can try to estimate the larger exponents by tracking the evolution of small sample subspaces through the tangent space.

These methods are described in detail in a review article by Abarbanel and coworkers [1993].

4.2.9 Lyapunov (Kaplan–Yorke) Dimension

We present below a heuristic discussion that motivates a connection between Lyapunov exponents and dimension. A schematic illustration of the heuristic argument for the Lyapunov dimension is shown in Fig. 4.5. Consider a two-dimensional map. Suppose we wish to compute the capacity of a chaotic attractor for which $\lambda_1 > 1 > \lambda_2$. Cover the attractor with $N(\varepsilon)$squares of side ε. Now, iterate the map q times. For a fixed q and ε small enough, the action of the mapping is roughly linear over the square, and each square will be stretched into a long thin parallelogram. From the definition of Lyapunov numbers, the average length of these parallelograms is $(\lambda_1)^{q_\varepsilon}$, and the average width is $(\lambda_2)^{q_\varepsilon}$. Now, suppose we had used a finer cover of squares of side $(\lambda_2)^{q_\varepsilon}$ (see the figure). To cover each parallelogram takes about $(\lambda_1/\lambda_2)^q$ smaller squares. Thus, *if it is supposed* that all squares on the attractor behave in this typical way, then we are led to the estimate

$$N(\lambda_2^q \varepsilon) \approx \left(\frac{\lambda_1}{\lambda_2}\right)^q N(\varepsilon) \qquad (4.20)$$

Motivated by Eq. (4.5), assume $N(\varepsilon) \approx k(1/\varepsilon)^{D_0}$, and substitute it into both sides of Eq. (4.20). This gives

$$k\left(\frac{1}{\lambda_2^q \varepsilon}\right)^{D_0} \approx k\left(\frac{\lambda_1}{\lambda_2}\right)^q \left(\frac{1}{\varepsilon}\right)^{D_0} \qquad (4.21)$$

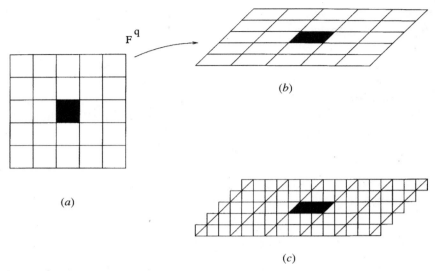

Figure 4.5 Schematic illustration of heuristic argument for Lyapunov dimension. Image of each square in (*a*) is approximately a parellelogram that has been stretched horizontally by a factor of λ_1^q and contracted vertically by a factor λ_2^q. The images in (*b*) thus have a smaller cover of squares as shown in (*c*).

Collecting terms, taking logarithms, and solving for D_0, we get

$$D_0 = 1 + \frac{\log \lambda_1}{\log(1/\lambda_2)} \qquad (4.22)$$

Farmer et al. [1983] showed that this expression is often meaningful even when the heuristic derivation was invalid and hence called it the *Lyapunov dimension*. This quantity was originally defined by Kaplan and Yorke [1979], who originally gave it as a lower bound on the fractal dimension. This dimension is also known as the *Kaplan–Yorke dimension* because of its originators.

The Kaplan–Yorke dimension gives a meaningful measure of the dimension of a strange attractor, defined in terms of its Lyapunov exponents [Kaplan and Yorke, 1979; Kaplan et al., 1983; Ledrappier, 1981; Russell et al., 1980; Young, 1982]. In order to formulate this measure, Kaplan and Yorke examined numerically a class of systems of the form

$$x(n) = f(x(n-1)) \qquad (4.23)$$

where $f: D \rightarrow D$, D is a compact n-dimensional set. If there exist n Lyapunov exponents $(\lambda_2, \lambda_2, \ldots, \lambda_n)$, then the dimension of the attractor is related to the Lyapunov exponents in the following way:

$$D_{KY} = K + \frac{\sum_{a=1}^{K} \lambda_a}{|\lambda_{K+1}|} \qquad (4.24)$$

where $\lambda_1 \geq \lambda_2 \geq \cdots \geq \lambda_K \geq 0$, and $\sum_{a=1}^{K} \lambda_a > 0$, $\sum_{a=1}^{K+1} \lambda_a < 0$. Clearly, $K < n$.

Although Kaplan and Yorke conjectured that $D_2 = D_{KY}$, in reality they need not be equal because of the fact that the estimation of D_{KY} uses only the "core" of the attractor where the trajectories of the attractor reside most of the time.[2] The fractal dimension, in contrast, is used for measuring the point densities of the entire attractor. Thus, Lyapunov exponents provide an important link between the fractal geometry of the attractor and the property of sensitive dependence on initial conditions.

4.2.10 Kolmogorov Entropy (KE)

The *Kolmogorov entropy*, also known as metric entropy or Kolmogorov–Sinai entropy, is the most important measure by which a chaotic motion in an arbitrary dimensional phase space can be characterized [Kolmogorov, 1958; Sinai, 1959, 1976]. The Kolmogorov entropy of an attractor can be considered as a measure of the rate of information loss along the attractor or as a measure of the degree of predictability of points along the attractor, given an arbitrary initial point. The above definition is made clear in the following sense. Due to the sensitivity to initial conditions in chaotic systems, nearby orbits diverge. If we can only distinguish orbit locations in phase space to within some given accuracy, then the initial conditions for two orbits may appear to be the same. As their orbits evolve forward in time, however, they will eventually move far enough apart that they may be distinguished as different. Alternatively, as an orbit is iterated, by observing its location with the given accuracy that we have, initially insignificant digits in the specification of the initial condition will even-

---

[2] There are some discrepancies regarding whether D_{KY} is an upper bound or lower bound to D_2 as reported in the literature. Kaplan and Yorke [1979] originally conjectured that they should be equal, but later Young [1982, 1983] reported that D_{KY} is a lower bound to D_2. Later again, Grassberger and Procaccia [1983a,b], suggested that D_{KY} is an upper bound to D_2.

tually make themselves felt. Thus, assuming that we can calculate exactly and that we know the dynamical equations giving an orbit, if we view that orbit with limited precision, we can, in principle, use our observation to obtain more and more information about the initial unresolved digits specifying the initial condition. It is in this sense that we say that a chaotic orbit creates information.

The KE can be calculated as follows [Farmer, 1982a,b]: Consider the trajectory of a dynamical system on a strange attractor. Divide the phase space into D-dimensional hypercubes of volume ε^D. Let $P_{i_0, i_1, \ldots i_n}$ be the probability that a trajectory is in hypercube i_0 at $t = 0$, i_1 at $t = T$, i_2 at $t = 2T$, and so on. Then the quantity

$$K_n = h_K = - \sum_{i_0, i_1, \ldots, i_n} P_{i_0, i_1, \ldots, i_n} \ln P_{i_0, i_1, \ldots, i_n} \tag{4.25}$$

is proportional to the information needed to locate the system on a special trajectory with precision ε. Therefore, $K_{N+1} - K_N$ is the additional information needed to predict which cube the trajectory will be at $(n + 1)T$ given trajectories up to nT. This means that $K_{N+1} - K_N$ measures our loss of information about the system from time n to time $n + 1$. The Kolmogorov entropy is then defined as the average rate of loss of information in the following way:

$$K \equiv \lim_{T \to 0} \lim_{\varepsilon \to 0^+} \lim_{N \to \infty} \frac{1}{NT} \sum_{n=0}^{N-1} (K_{n+1} - K_n) \tag{4.26}$$

The order in which the limits in the above expression are taken is immaterial. The limit $\varepsilon \to 0$ (which has to be taken *after $N \to \infty$*) makes KE independent of the particular partition. The main properties of KE are as follows:

1. The entropy K determines the rate of change in entropy as a result of a purely dynamic process of mixing of trajectories in phase space.
2. The entropy, increment of local instability, and inverse decay correlation time are quantities of the same order of magnitude
3. The entropy is a metric invariant of the system; that is, its value is independent of the way the phase space is divided into cells and coarsened.
4. Systems with identical values of entropy are in a certain sense isomorphic to each other [Billingsley, 1965; Ornsteien, 1974]; that is, these systems must have identical statistical laws of motion.

4.2.11 Relationship between the Kolmogorov Entropy and Lyapunov Exponents

The connection between the KE and the Lyapunov exponents is well understood. In one-dimensional maps, KE is just the Lyapunov exponent [Schouten, 1994b]. In higher dimensional systems, we lose information about the system because the cell in which it was previously located spreads over new cells in phase space at a rate determined by the Lyapunov exponents. The rate K at which the information about the system is lost is equal to the (averaged) sum of positive Lyapunov exponents [Pesin, 1977], as shown by

$$K = \sum_i \lambda_i \qquad (4.27)$$

where λ_i's are the positive Lyapunov exponents of the dynamical system under study.

4.2.12 Embedding Dimension (d_E)

The minimum number of elements in the time-delay embedding process that achieves the dynamic reconstruction of the underlying dynamic process is called the *embedding dimension*. It is denoted by d_E throughout this book. The procedure of finding a suitable value of d_E is called *embedding*. The embedding dimension d_E is a global dimension. In general, it is different from the *local dimension d_L* that defines the number of true Lyapunov exponents of the system under investigation.

4.2.13 Embedding Delay (τ)

In time-delay embedding, the time lag between two consecutive elements of the embedding vector (a point in the state space) is called the *embedding delay*. It is denoted by τ. It should be large enough so that $y(n)$ and $y(n - \tau)$ are essentially independent of each other so as to serve as independent coordinates of the reconstruction space but at the same time not so independent as to have no correlation with each other. This requirement is best satisfied by choosing a delay τ for which the mutual information between $y(n)$ and $y(n - \tau)$ attains its first minimum [Fraser, 1989; Fraser and Swinney, 1986].

4.3 CRITERIA FOR ASSESSING THE CHAOTIC DYNAMICS OF AN EXPERIMENTAL TIME SERIES

Given a physical (experimental) time series, how do we determine whether the time series is derived from a chaotic process? Unfortunately, there is no single criterion that we may invoke to answer this important question. Rather, we have a set of criteria, all of which must be satisfied if we are to be certain that the given time series is the result of a chaotic process. These criteria are summarized here:

1. The process should be nonlinear.
2. The attractor correlation dimension (D_2) of the process should be fractal. Moreover, D_2 should converge to a constant value for increasing embedding dimension.
3. The dynamics of the system responsible for the generation of the process should be sensitive to initial conditions. This, in turn, means that at least one of the Lyapunov exponents of the process should be positive. The largest Lyapunov exponent determines the horizon within which the time series is predictable.
4. The sum of all Lyapunov exponents of the process should be negative for the underlying dynamics to be dissipative (i.e., physically realizable).
5. The Kaplan–Yorke dimension (D_{KY}), a by-product of the Lyapunov spectrum, should be close in numerical value to the correlation dimension.
6. The Kolmogorov entropy, or metric entropy, of the system should be positive and finite. Its numerical value, determined by a method of its own, should be close to the sum of all positive Lyapunov exponents.
7. The local embedding dimension $d_L \leq d_E$, where d_E is the global embedding dimension.

5

PREPROCESSING OF THE RADAR DATA

5.1 INTRODUCTION

Preprocessing of the measured data is an essential step in any experimental data analysis. The experimental radar data collected from different sites using different radars were influenced by many types of errors: (i) receiver noise, (ii) quantization error due to the A/D converter, and (iii) in-phase and quadrature-phase component misalignment due to gain and phase imbalances. These errors were minimized by suitable preprocessing of the data before proceeding to extract the chaotic parameters. Most importantly, however, extra care was taken to ensure that the preprocessing did not alter the dynamics of the underlying clutter process. In this study, the following preprocessing techniques were employed on the radar data in the order presented here:

- Amplitude and phase correction (I–Q calibration)
- Filtering
- Continuity and differentiability tests on filtered data

5.2 AMPLITUDE AND PHASE CORRECTIONS (I–Q CALIBRATION)

The raw in-phase (I) and quadrature-phase (Q) components of the coherent radar showed amplitude and phase mismatches (unequal variance and nonzero cross-correlation) because of experimental errors. These were due to direct current (dc) offsets and differences in the gain and phase characteristics of the individual receivers. These errors were corrected by the I–Q calibration method, as detailed below.

The I–Q calibration procedure was applied to measure the gain and phase imbalances between the I- and Q-channels. Figure 5.1 shows the circuit diagram for doing this calibration. The low-pass filters (LPFs) included in this figure were designed to remove the high-frequency components appearing at the multiplier outputs. The receiver channel gains for both I- and Q-channels were normalized by using the I-channel as reference. This resulted in a unity gain for the I-channel and G_e for the Q-channel. The symbol θ_e represents the relative phase error in the Q-channel with respect to the I-channel. The terms Dc_I and Dc_Q represent

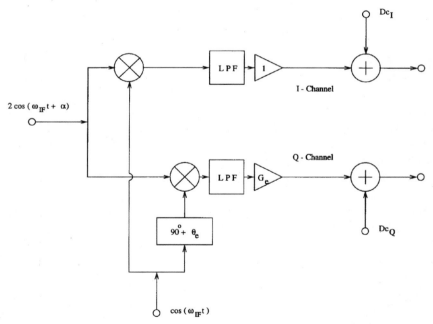

Figure 5.1 Circuit model for I–Q calibration.

the dc offsets in I- and Q-channels, respectively. If the input is an IF signal of the form $2\cos(\omega_{IF}t + \alpha)$, the outputs of the I- and Q-channels would ideally be as follows (assuming no errors):

$$I = A\cos(\alpha) \tag{5.1}$$

and

$$Q = A\sin(\alpha) \tag{5.2}$$

where A is a scaling factor. However, because of unavoidable gain and phase errors, as well as the dc offsets in the two channels, the actual outputs in Fig. 5.1 are

$$I' = A\cos(\alpha) + Dc_I \tag{5.3}$$

and

$$Q' = AG_e\sin(\alpha - \theta_e) + Dc_Q \tag{5.4}$$

Now, solving for $\cos(\alpha)$ and $\sin(\alpha)$ from Eqs. (5.3) and (5.4) and substituting them into Eqs. (5.1) and (5.2), we obtain the actual I and Q components of the signal as follows:

$$I = I' - Dc_I \tag{5.5}$$

and

$$Q = \frac{1}{\cos(\theta_e)}\left[\frac{Q' - Dc_Q}{G_e} - I\sin(\theta_e)\right] \tag{5.6}$$

The parameters G_e and θ_e for the recorded I- and Q-values (for a data length N) are computed using the following formulas:

$$G_e = \left(\frac{\sum_{i=1}^{N}[Q'(i) - Dc_Q]^2}{\sum_{i=1}^{N}[I'(i) - Dc_I]^2}\right)^{1/2} \tag{5.7}$$

and

$$\theta_e = \sin^{-1}\left(\frac{2\sum_{i=1}^{N}[I'(i) - Dc_I]\{[Q'(i) - Dc_Q]/G_e\}}{\sum_{i=1}^{N}[I'(i) - Dc_I]^2 + \{[Q'(i) - Dc_Q]/G_e^2\}}\right) \tag{5.8}$$

This calibration process was applied to the sea clutter data from Cape Bonavista and Dartmouth, where I- and Q-channel data are available separately. In each case, the amplitude of the clutter data was computed in the following way (after performing filtering operations on I–Q corrected I and Q components, independently):

$$A = \sqrt{(I^2 + Q^2)} \tag{5.9}$$

As for the Argentia data sets, there is no need for such a correction to be made as the radar system has a single component, namely, amplitude.

5.2.1 Effect of Miscalibration

From our studies, we found that the I–Q calibration procedure is an important step in the preprocessing of the collected clutter data using a coherent radar.

In order to demonstrate the necessity of performing the I–Q calibration before extracting the chaotic parameters of the clutter data collected using a coherent radar, we extracted the chaotic parameters of the three-point smoothed in-phase, quadrature-phase, and amplitude components of clutter data (for different ranges) for data sets from the experiments at Cape Bonavista. These results were compared to those obtained on the same sets of clutter data after performing the I–Q calibration. It was observed that the chaotic invariants (D_2, D_{KY}, and λ_i) were different from each other. Table 5.1 shows the correlation dimension (maximum-likelihood method) and the Kaplan–Yorke dimension of three-point smoothed amplitude, in-phase, and quadrature-phase components of clutter data for ranges 6150, 6300, and 6450 of the Cape Bonavista database for cases before and after I–Q calibration. It may be observed that the estimates of both D_2 and D_{KY} give higher values in the case of miscalibrated data compared to the calibrated data sets. It may also be observed that the Kaplan–Yorke dimension is consistently higher compared to the calibrated data sets. In one particular case (range 6150, in-phase), for an embedding dimension $d_E = 5$, the algorithm did not even compute the Kaplan–Yorke dimension. Similar behavior was observed in other data sets.

Table 5.2 gives the comparison of the Lyapunov spectrum of the same data sets before and after I–Q calibration. It may be observed that the amplitudes of the positive exponents are consistently higher in the case of data sets before applying the I–Q calibration. It may also be observed that the Lyapunov spectrum for data sets before I–Q calibration do not consistently follow the pattern of two positive Lyapunov exponents, with the third one approaching zero, followed by two or three negative exponents that we observed in the I–Q calibrated data sets. The results

Table 5.1   Correlation Dimension of In-Phase, Quadrature Phase, and Amplitude of Three-Point Smoothed Clutter Data for Different Ranges (Bonavista)

	Before I–Q Calibration					
	Amplitude		I-Phase		Q-Phase	
Range	D_{KY}	D_{ML}	D_{KY}	D_{ML}	D_{KY}	D_{ML}
6150	4.83	4.10	—	4.38	4.68	4.44
6300	4.73	4.78	4.98	5.86	4.57	5.12
6450	4.45	4.28	4.50	4.62	4.45	4.54
	After I–Q Calibration					
6150	4.20	4.31	4.32	4.23	4.52	4.63
6300	4.17	4.21	4.30	4.14	4.21	4.38
6450	4.15	4.26	4.10	4.32	4.10	4.06

Table 5.2   Lyapunov Exponents of I-Phase, Q-Phase, and Amplitude Components of Three-Point Smoothed Clutter Data for Different Ranges (Bonavista)

	Before I–Q Calibration			After I–Q Calibration		
Range	Amplitude	I-Phase	Q-Phase	Amplitude	I-Phase	Q-Phase
6150	+0.4833	+0.5716	+0.4107	+0.2593	+0.2712	+0.3786
	+0.2971	+0.3897	+0.2433	+0.1244	+0.1493	+0.2149
	+0.0997	+0.1693	+0.0567	−0.0219	−0.0065	+0.0429
	−0.2100	−0.1723	−0.1968	−0.2275	−0.1924	−0.2310
	−0.8073	−0.8421	−0.7512	−0.7168	−0.6921	−0.7797
6300	+0.4217	+0.5111	+0.3827	+0.2676	+0.2532	+0.2658
	+0.2491	+0.3318	+0.2219	+0.1382	+0.1441	+0.1251
	+0.0897	+0.1178	+0.0310	−0.0143	−0.0056	−0.0134
	−0.2100	−0.1921	−0.1993	−0.2573	−0.1914	−0.2308
	−0.7931	−0.7917	−0.7817	−0.8092	−0.6604	−0.7071
6450	+0.3211	+0.3171	+0.3122	+0.2550	+0.1670	+0.1610
	+0.1871	+0.1870	+0.1712	+0.1347	+0.0704	+0.0660
	+0.0293	+0.0317	+0.0210	−0.0209	−0.0296	−0.0324
	−0.1897	−0.1879	−0.1792	−0.2558	−0.1702	−0.1505
	−0.7219	−0.6810	−0.7124	−0.7493	−0.5251	−0.5450

presented here clearly show that the I–Q calibration of the radar data is an important preprocessing step in the case of clutter data from a coherent radar if the results of the chaotic analysis are to be trusted.

5.3   FILTERING

The collected data may also be influenced by measurement noise (quantization error and receiver noise) or additive noise (artifacts). These artifacts must be minimized by applying suitable filters to the recorded signal. Here, again, care must be taken in choosing the smoothing filters to make sure that the filtering does not affect the underlying dynamics of the system that produced the time series [Badii et al., 1988; Broomhead et al., 1992; Mitschke, 1990; Paoli et al., 1989; Pecora and Carrol, 1996; Rapp et al., 1993]. We used two kinds of filtering to reduce the effect of experimental noise: three-point smoothing and finite impulse response (FIR) filtering.

5.3.1   Three-Point Smoothing

For discrete-time sequences a common smoothing operation is one referred to as a moving average, where the smoothed value $\hat{x}(n)$ for any n is an average of values of $x(n)$ in the vicinity of n. The basic idea is that by averaging values locally, rapid variations from point to point will be averaged out and slow variations will be retained, corresponding to smoothing or low-pass filtering the original sequence. In the first method, we used a smoothing filter of the form

$$\hat{x}(n) = \sum_{i=1}^{3} a_i x(n + i - 2) \tag{5.10}$$

with $a_i = \frac{1}{3}$. Here, the output is the average of three consecutive input values. Equation (5.10) is a nonrecursive linear constant-coefficient difference equation. The filter transfer function is given by

$$H(\omega) = \frac{1}{3}(e^{j\omega} + 1 + e^{-j\omega})$$
$$= \frac{1}{3}[1 + 2\cos(\omega)] \tag{5.11}$$

Hence, the magnitude and phase responses of the filter are given by

$$|H(\omega)| = \frac{1}{3}|1 + 2\cos(\omega)| \tag{5.12}$$

and

$$\theta(\omega) = \begin{cases} 0 & 0 \le \omega \le \frac{2}{3}\pi \\ \pi & \frac{2}{3}\pi \le \omega < \pi \end{cases} \qquad (5.13)$$

The magnitude and phase responses of this filter are shown in Figs. 5.2a and 5.2b, respectively. We observe that it has the general character-

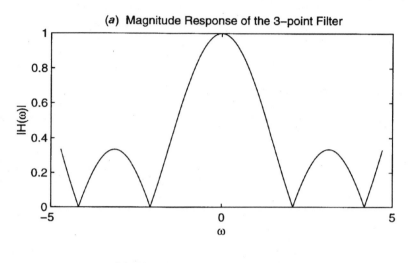

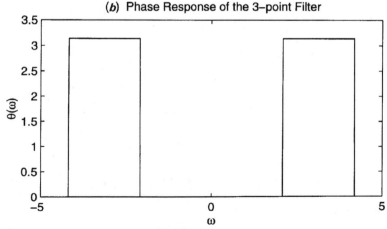

Figure 5.2   Frequency responses of three-point smoothing filter: (a) magnitude response; (b) phase response.

istics of a low-pass filter, although it does not have a sharp transition from passband to stopband.

5.3.2 FIR Filtering

In the second method, we designed a low-pass linear phase, finite impulse response (FIR) filter having 101 coefficients to filter the raw clutter data. While designing the filter, care was taken to ensure that the bandwidth of the filter was large enough so as not to affect the clutter data. This issue is discussed in the next section. The FIR filter coefficients are tabulated in Table 5.3. The magnitude response of the filter is given by

$$H(\omega) = H_r(\omega)e^{-j\omega(M-1)/2} \tag{5.14}$$

Table 5.3   FIR Filter Coefficients

$h(0) = +5.0905235 \times 10^{-4} = h(100)$	$h(26) = -5.5413651 \times 10^{-18} = h(74)$
$h(1) = +3.7146788 \times 10^{-4} = h(99)$	$h(27) = -5.8458671 \times 10^{-3} = h(73)$
$h(2) = -8.1459374 \times 10^{-19} = h(98)$	$h(28) = -9.0558591 \times 10^{-3} = h(72)$
$h(3) = -4.2193097 \times 10^{-4} = h(97)$	$h(29) = -7.0105114 \times 10^{-3} = h(71)$
$h(4) = -6.5327286 \times 10^{-4} = h(96)$	$h(30) = +6.6446455 \times 10^{-18} = h(70)$
$h(5) = -5.1250504 \times 10^{-4} = h(95)$	$h(31) = +8.3989821 \times 10^{-3} = h(69)$
$h(6) = +3.9788816 \times 10^{-18} = h(94)$	$h(32) = +1.3006614 \times 10^{-2} = h(68)$
$h(7) = +6.4760162 \times 10^{-4} = h(93)$	$h(33) = +1.0078811 \times 10^{-2} = h(67)$
$h(8) = +1.0370345 \times 10^{-3} = h(92)$	$h(34) = -7.6609245 \times 10^{-18} = h(66)$
$h(9) = +8.3189599 \times 10^{-4} = h(91)$	$h(35) = -1.2154177 \times 10^{-2} = h(65)$
$h(10) = -1.6350096 \times 10^{-18} = h(90)$	$h(36) = -1.8935271 \times 10^{-2} = h(64)$
$h(11) = -1.0704253 \times 10^{-3} = h(89)$	$h(37) = -1.4794326 \times 10^{-2} = h(63)$
$h(12) = -1.7136513 \times 10^{-3} = h(88)$	$h(38) = +8.5263457 \times 10^{-18} = h(62)$
$h(13) = -1.3687285 \times 10^{-3} = h(87)$	$h(39) = +1.8292965 \times 10^{-2} = h(61)$
$h(14) = +2.4038758 \times 10^{-18} = h(86)$	$h(40) = +2.9020687 \times 10^{-2} = h(60)$
$h(15) = +1.7330439 \times 10^{-3} = h(85)$	$h(41) = +2.3206829 \times 10^{-2} = h(59)$
$h(16) = +2.7466338 \times 10^{-3} = h(84)$	$h(42) = -9.1865313 \times 10^{-18} = h(58)$
$h(17) = +2.1705883 \times 10^{-3} = h(83)$	$h(43) = -3.0731761 \times 10^{-2} = h(57)$
$h(18) = -3.3522037 \times 10^{-18} = h(82)$	$h(44) = -5.1313388 \times 10^{-2} = h(56)$
$h(19) = -2.6899528 \times 10^{-3} = h(81)$	$h(45) = -4.3981294 \times 10^{-2} = h(55)$
$h(20) = -4.2193247 \times 10^{-3} = h(80)$	$h(46) = +9.5999996 \times 10^{-18} = h(54)$
$h(21) = -3.3016710 \times 10^{-3} = h(79)$	$h(47) = +7.4379480 \times 10^{-2} = h(53)$
$h(22) = +4.4204065 \times 10^{-18} = h(78)$	$h(48) = +1.5850184 \times 10^{-1} = h(52)$
$h(23) = +4.0190464 \times 10^{-3} = h(77)$	$h(49) = +2.2476727 \times 10^{-1} = h(51)$
$h(24) = +6.2544467 \times 10^{-3} = h(76)$	$h(50) = +2.4988049 \times 10^{-1} = h(50)$
$h(25) = +4.8593839 \times 10^{-3} = h(75)$	

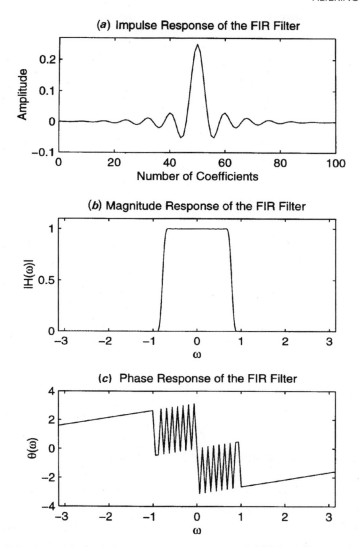

Figure 5.3 Impulse and frequency responses of 101-tap linear phase FIR filter: (*a*) impulse response; (*b*) magnitude response; (*c*) phase response.

where $H_r(\omega)$ is a real function of ω and can be expressed as

$$H_r(\omega) = h\left(\frac{M-1}{2}\right) + 2 \sum_{n=0}^{(M-3)/2} h(n) \cos\left[\omega\left(\frac{M-1}{2} - n\right)\right] \quad (M \text{ odd})$$

$$(5.15)$$

$$H_r(\omega) = 2 \sum_{n=0}^{M/2-1} h(n) \cos\left[\omega\left(\frac{M-1}{2} - n\right)\right] \quad (M \text{ even}) \quad (5.16)$$

The phase characteristic of the filter for both M odd and M even is

$$\theta(\omega) = \begin{cases} -\omega[\frac{1}{2}(M-1)] & \text{if } H_r(\omega) > 0 \\ -\omega[\frac{1}{2}(M-1)] + \pi & \text{if } H_r(\omega) < 0 \end{cases} \quad (5.17)$$

The filter impulse response and the magnitude response are shown in Figs. 5.3a and 5.3b, respectively. The phase response is plotted in Fig. 5.3c. The frequency axis in Fig. 5.3b is normalized to the pulse repetition frequency of the radar. It may be observed that the filter magnitude response is flat in the passband and the stopband response is essentially zero. This characteristic of the filter ensures that the filtering operation attenuates the out-of-band signals (noise) very effectively without affecting signals in the band of interest at all.

5.4   RESULTS OF PREPROCESSING OF THE SIGNALS

Each sea clutter data set had at least 50,000 samples taken from a fixed resolution cell on successive sweeps. The sampling period is determined by the pulse repetition interval of the radar. In each case, the data were preprocessed for amplitude and phase correction (coherent radar data), as described in Section 5.2. Once the I–Q calibration was done, the data (I and Q components) were subjected to filtering (three-point smoothing or FIR filtering) before computing the amplitude (in the case of coherent radar data) of the clutter data. In the case of noncoherent radar data sets, the amplitude of the clutter data is available in the first place, and the filtering is performed on the amplitude data alone. Figure 5.4 illustrates the unfiltered and filtered amplitude components of sea clutter data for

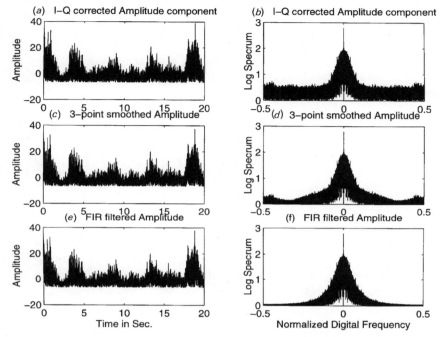

Figure 5.4 Effect of filtering: (*a*) I–Q corrected amplitude component of sea clutter; (*b*) corresponding FFT spectrum; (*c*) three-point smoothed amplitude; (*d*) corresponding spectrum; (*e*) FIR filtered amplitude and (*f*) corresponding spectrum.

20 s duration; Fig. 5.4*a* refers to unfiltered data, and Figs. 5.4*c* and 5.4*e* refer to filtered data, using the three-point smoother and FIR filter, respectively. The corresponding fast Fourier transform (FFT) spectra are shown in Figs. 5.4*b*, 5.4*d*, and 5.4*f*, respectively. We observe that the unfiltered amplitude signal has high-frequency noise components, as its FFT spectrum clearly indicates. Figure 5.4*d*, which is the FFT spectrum of the three-point smoothed amplitude component, clearly shows the attenuation of the high-frequency components without affecting the actual clutter spectrum. The attenuation of noise is almost complete in the second filtered case shown in Fig. 5.4*f*, where we used a 101-tap FIR filter for suppressing the high-frequency noise from the amplitude component. Here, again, the filtering did not alter the spectrum of the sea clutter. After these preprocessing procedures were completed, each data set was

analyzed to test whether the preprocessing had altered the underlying dynamics [Badii et al., 1988; Broomhead et al., 1992; Mitschke, 1990; Paoli et al., 1989; Pecora and Carrol, 1996; Rapp et al., 1993]. This was done by statistically testing for the continuity and differentiability between the original and preprocessed data embedding, as described in the next section.

5.5 CONTINUITY AND DIFFERENTIABILITY TESTS ON FILTERED DATA

We apply a set of statistics intended to characterize in terms of probabilities or confidence levels whether two data sets are related by mapping with certain mathematical properties. Here, the two data sets are the unfiltered and filtered versions of the time series and the mapping is the filtering operation. We consider the mathematical properties of continuity and differentiability between these two data sets to test whether the mapping has really preserved its characteristics to each other.

The effect of preprocessing on the collected data is tested by computing the continuity and differentiability of these two (raw and preprocessed) data sets. We are interested in characterizing, in a statistical sense, the mathematical properties of embedding from one time series to a second time series. By embedding, we mean the reconstruction of an unknown dynamical system by formulating a pseudo–phase space using the successive values of the observable time series. Figure 5.5 is a diagram of the mapping between two time series phase spaces. In our case, the first time series is the unfiltered sea clutter data and the second time series is the filtered sea clutter data. The essence of this mathematical analysis is to show that the trajectories formed from these two time series are diffeomorphically related to each other. *Diffeomorphism* means that the map between the original and the filtered signals has no tears (is continuous), does not glue together geometrically separate points, and is smooth from the phase space of the original time series to the phase space of the filtered time series (differentiable) and in the reverse direction (differentiable inverse). The diffeomorphic relationship between these two time series is analyzed by testing for *continuity* and *differentiability* [Pecora et al., 1995], which are important properties pertaining to the preservation of the link between dynamics of the two time series.

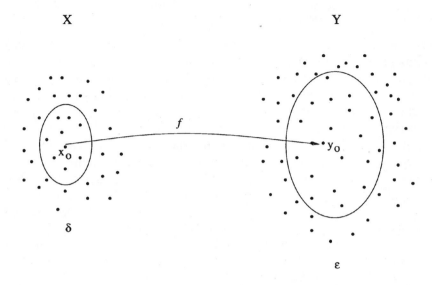

Figure 5.5 Illustration of mapping of one phase space to another.

5.6 CONTINUITY

5.6.1 Definition

Let **X** be the phase space constituted by the first time series and **Y** be the phase space constituted by the second time series. Let f be the mapping (linear filtering in our case) that generates the second time series from the first. The map f is said to be *continuous* if there are no points in the phase space **X** that are arbitrarily close but that are mapped to distant points on the phase space **Y**. Putting it in a mathematical form, the function f is continuous if the following condition is satisfied:

At a point $\mathbf{x}_0 \in \mathbf{X}$ if $\forall \varepsilon > 0 \; \exists \delta > 0$ such that $\|\mathbf{x} - \mathbf{x}_0\| < \delta \Rightarrow \|f(\mathbf{x}) - f(\mathbf{x}_0)\| < \varepsilon$.

This means that if we pick an open ε-sized set around $f(\mathbf{x}_0)$, then there is some small enough δ-sized set around $\mathbf{x}_0$ from which *all* of the points are mapped by f into the ε set. This guarantees that nearby points are not mapped to distant points.

5.6.2 Test of Continuity

The continuity is tested in the following way. When we only know some points in the domain and range of f, but not f itself, the statistical version of continuity can be derived by counting the points in the δ and ε sets. The statistical version can be derived in two steps. In the first step, we construct an algorithm to select points that, given δ and ε, are consistent with the mathematical definition of continuity. In the second step, we apply an appropriate probability distribution consistent with the null hypothesis we have chosen. These two steps are independent.

For the first step (the algorithm for finding points), start with ε and δ values. Find all the points $\mathbf{x}_i$ that are within a distance (Euclidean) δ of $\mathbf{x}_0$. Now, check to see if *all* the images $\mathbf{y}_i = f(\mathbf{x}_i)$ of those points are within a distance ε of $\mathbf{y}_0 = f(\mathbf{x}_0)$. If not, decrease δ by some factor, find a new set of $\mathbf{x}_i$'s, and check that the new images are within ε of $\mathbf{y}_0$. Repeat this process until we either have all $\mathbf{y}_i$'s in the ε set or δ has been decreased so far that we cannot find any $\mathbf{x}_i$ points other than $\mathbf{x}_0$.

For the second step, we introduce a null hypothesis that we use to generate a probability for our simple statistic for continuity. Let there be N points in the reconstructed embedding space. Let n_δ be the number of $\mathbf{x}_i$ points found in the δ set and n_ε be the total number of $\mathbf{y}_i$ points found in the ε set. Define a null hypothesis that the $\mathbf{y}$ points are randomly and independently distributed over the $\mathbf{Y}$ space reconstruction with respect to the $\mathbf{x}$ points. For this situation, the probability p of one of the n_δ points mapping into the ε set is

$$p = \frac{n_\varepsilon}{N} \qquad (5.18)$$

The probability of n_δ $\mathbf{x}$ points mapping into the ε set is

$$p^{n_\delta} = \left[\frac{n_\varepsilon}{N}\right]^{n_\delta} \qquad (5.19)$$

If this probability is low enough, then we can reject the null hypothesis; the $\mathbf{x}$ and $\mathbf{y}$ points are not randomly related. This does not mean that they are related in a particular way. Rejection of the null hypothesis does not imply an automatic acceptance of the alternate hypothesis that the mapping is continuous [Arnold, 1981; Seber, 1984]. This means the fact that in this case n_δ points all mapped into the ε set is not caused by some random coincidence.

We express this confidence by quantifying what we meant above by having a probability that is "low enough." First of all, it is obvious that we are dealing with a binomial distribution $b(m; n_\delta, p)$, which gives the probability of finding m points out of n_δ inside the ε set if the probability for finding one point in the ε set is p [Walpole and Myers, 1972]. In our case, the events are in the "tail" of the distribution, $m = n_\delta$, and so the probability of this happening is p^{n_δ}. The likelihood of this happening is defined as the ratio of probabilities $p^{n_\delta}/p_{\max}$. The quantity $p_{\max}$ is the maximum in the binomial distribution that occurs, usually, at some intermediate $m < n_\delta$ value. This implies that it is not enough to have the probability p^{n_δ} of the event be small, but it must be small compared to the probability of the most likely event. This probability is denoted by $p_{\max}$.

Now, define the confidence as the continuity index $\Theta_{C^0}(\varepsilon)$ in the following way:

$$\Theta_{C^0}(\varepsilon) = \frac{1}{n_p} \sum_{j=1}^{n_p} \Theta_{C^0}(\varepsilon, j) \tag{5.20}$$

where

$$\Theta_{C^0}(\varepsilon, j) = 1 - \frac{p^{n_\delta}}{p_{\max}} \tag{5.21}$$

and j is the index of the point at which we are testing the continuity $(\mathbf{x}_0 = \mathbf{x}_j)$; n_p is the number of centers chosen randomly.

When $\Theta_{C^0}(\varepsilon) \approx 1$, we can be confident that the mapping is continuous (by rejecting the null hypothesis).

5.7   DIFFERENTIABILITY

5.7.1   Definition

This property tells us about the smoothness of the function f. A mathematical definition for differentiability is as follows:

The mapping $f : \mathbf{X} \to \mathbf{Y}$ is said to be differentiable at a point $\mathbf{x}_0 \in \mathbf{X}$ if there exists a linear operator $\mathbf{A}$ such that $\forall \varepsilon > 0 \; \exists \delta > 0$ for which $\|\mathbf{x} - \mathbf{x}_0\| < \delta \Rightarrow \|f(\mathbf{x}_0) + \mathbf{A}(\mathbf{x} - \mathbf{x}_0) - f(\mathbf{x})\| < \varepsilon \|\mathbf{x} - \mathbf{x}_0\|$, where ε is a scaling factor.

This means that f is well approximated locally by $\mathbf{A}$ and is therefore smooth. We have chosen to use the well-known statistic of multivariate correlation and use a null hypothesis that generates a probability for confidence statistic analogous to that for the continuity at $\mathbf{x}_0$. Here, we define another quantity called the *differentiability index*, denoted by $\Theta_{C^1}(\varepsilon)$, to statistically test for smoothness of the mapping. When $\Theta_{C^1}(\varepsilon) \approx 1$, we can be confident that the mapping is differentiable (by rejecting the null hypothesis).

5.7.2 Test of Differentiability

We have chosen to use the well-known statistic of multivariate correlation and use a null hypothesis that generates a probability for confidence statistic analogous to that for the continuity at $\mathbf{x}_0$. It is known that the method of least squares yields the best estimate of a linear operator that satisfies the equation

$$\mathbf{Aa} = \mathbf{b} \tag{5.22}$$

for a given set of $(\mathbf{a}_i, \mathbf{b}_i)$ pairs of measurements when the data have a normal distribution. For our study, we use the $(\mathbf{a}_i, \mathbf{b}_i)$ pairs as the zero-mean variables $\Delta\mathbf{x}_i = \mathbf{x}_i - \bar{\mathbf{x}}$ and $\Delta\mathbf{y}_i = \mathbf{y}_i - \bar{\mathbf{y}}$, where $\bar{\mathbf{x}}$ is the mean of the n_δ vectors found for the continuity statistic and $\bar{\mathbf{y}}$ is their image. The least-squares solution is given by

$$\mathbf{A} = \mathbf{Y}\mathbf{X}^{\mathrm{T}}(\mathbf{X}\mathbf{X}^{\mathrm{T}})^{-1} \tag{5.23}$$

where $\mathbf{X}$ is the matrix whose columns are the $\mathbf{x}$ vectors, $\mathbf{X} = (\Delta\mathbf{x}_1, \Delta\mathbf{x}_2, \ldots, \Delta\mathbf{x}_n)$, and $\mathbf{Y}$ is the matrix whose columns are the $\mathbf{y}$ vectors, $\mathbf{Y} = (\Delta\mathbf{y}_1, \Delta\mathbf{y}_2, \ldots, \Delta\mathbf{y}_n)$. For the inverse of $\mathbf{A}$, we get the estimate

$$\mathbf{A}^{-1} = \mathbf{X}\mathbf{Y}^{\mathrm{T}}(\mathbf{Y}\mathbf{Y}^{\mathrm{T}})^{-1} \tag{5.24}$$

If we have good estimates of $\mathbf{A}$ and $\mathbf{A}^{-1}$, we should get $\mathbf{A}\mathbf{A}^{-1} = \mathbf{I}$ (the unit matrix). Putting this relation together with (5.23) and (5.24), the correlation statistic r is given by the relation

$$r^2 = \mathrm{tr}\left[\frac{\mathbf{X}\mathbf{Y}^{\mathrm{T}}(\mathbf{Y}\mathbf{Y}^{\mathrm{T}})^{-1}\mathbf{Y}\mathbf{X}^{\mathrm{T}}(\mathbf{X}\mathbf{X}^{\mathrm{T}})^{-1}}{d_a}\right] \tag{5.25}$$

where tr is the trace operator and d_a is the dimension of the attractor. This is typically the statistic used in multivariate analysis to estimate the amount of linear correlation between two data sets. Now, we ask how probable this correlation is for the null hypothesis. (The null hypothesis is that the Δx and Δy pairs are uncorrelated.)

The approximate probability for finding a correlation r^2 in spaces of d-dimensional vector pairs on a d_a-dimensional subspace that are assumed uncorrelated is

$$p \sim e^{-(1/2)(n_\delta - d_a - 1)^2 r^2 d_a} \tag{5.26}$$

As in the continuity index estimate, the process of determining a likelihood or confidence level is a two-step process. In the first step, we define our algorithm for finding points x_j that satisfy the definition of differentiability for the map (locally at x_0). We choose the ε-value, which in this case now determines the error we will allow in the local linear estimate of our function. This ε has a different meaning from that we have used in the continuity test. We pick a δ-value and find associated x_j points within δ of x_0. This gives us a set of local Δx and corresponding Δy pairs. We use these to find a least-squares approximation to A—the local linear map. Now, we check whether $\|\Delta y - A\,\Delta x\| < \varepsilon_s\,\Delta x$ as the definition of differentiability requires. We use a scaled $\varepsilon_s = \varepsilon(\sigma_{\Delta y}/\sigma_{\Delta x})$, where $\sigma_{\Delta y}$ is the standard deviation of the Δy vectors and $\sigma_{\Delta x}$ is the standard deviation of the Δx vectors. The use of this scaled ε_s is necessary to eliminate the scale difference between X and Y vectors. If all Δx and Δy pairs satisfy the inequality, we move on to the calculation of the statistic. If not, we choose a smaller δ and try again with a new and smaller set of Δx and Δy pairs. We continue this process until the inequality is satisfied or we run out of points.

The second step is to calculate the correlation r^2 and the associated probability p for obtaining r^2 given by Eq. (5.26), which is determined by the null hypothesis of no correlation between Δx and Δy pairs. If the probability p is low enough, then we can reject the null hypothesis; the x and y points are not randomly related. Now, define the average statistics $\Theta_{C^1}(\varepsilon)$ as

$$\Theta_{C^1}(\varepsilon) = \frac{1}{n_p} \sum_{j=1}^{n_p} \Theta_{C^1}(\varepsilon, j) \tag{5.27}$$

where

$$\Theta_{C^1}(\varepsilon,\ j) = 1 - p \tag{5.28}$$

and j is the index of the point at which we are testing the differentiability ($\mathbf{x}_0 = \mathbf{x}_j$). The number of centers n_p is chosen randomly.

When $\Theta_{C^1}(\varepsilon) \approx 1$, we can be confident that the mapping is differentiable (by rejecting the null hypothesis).

5.8   RESULTS OF CONTINUITY AND DIFFERENTIABILITY TESTS

Figure 5.6 shows the continuity statistic [Pecora and Carrol, 1996] $\Theta_{C^0}(\varepsilon)$ for one typical clutter (amplitude before and after filtering) time series of various lengths (1000, 10,000, and 50,000). We have chosen an embedding dimension $d_E = 5$, embedding delay $\tau = 11$, number of randomly chosen centers 50, and ε values ranging from 0.001 to 1.0. As expected, the continuity drops when the ε-value becomes smaller and smaller. This results from a simple lack of points in the small sets, so the null hypothesis cannot be rejected for few random centers around the attractor. However, when more points are added to the time series, confidence in continuity goes up at smaller ε-values also. When the mapping is continuous, the respective statistic continually improves and approaches 1.0 asymptotically. The continuity statistic $\Theta_{C^0}(\varepsilon)$ varies between 0.99 and 1.0. This is what we observe in Fig. 5.6.

Figure 5.7 illustrates the differentiability statistic [Pecora and Carrol, 1996] $\Theta_{C^1}(\varepsilon)$ for the same data set. Here, also, the differentiability statistic $\Theta_{C^1}(\varepsilon)$ approaches 1.0 asymptotically. The $\Theta_{C^1}(\varepsilon)$ varies between 0.93 and 1.0 (see Fig. 5.7). We extended these tests for several such clutter data sets, and a similar performance was observed. In all the cases, the filtered data passed the test, meaning that the underlying dynamics of experimental sea clutter data remained unchanged after the low-pass filtering to minimize the effect of additive receiver noise.

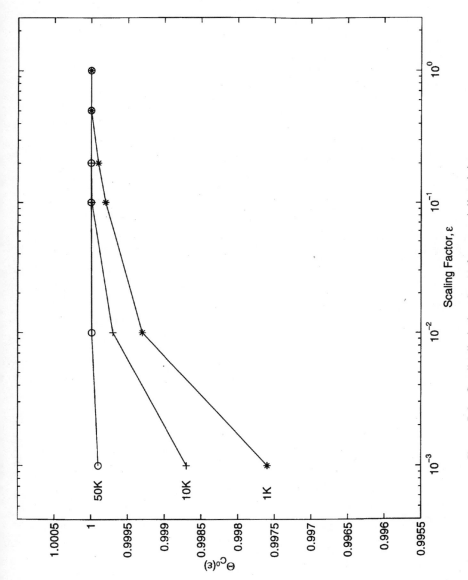

Figure 5.6 Continuity index $\Theta_{C^0}(\epsilon)$ for sea clutter data.

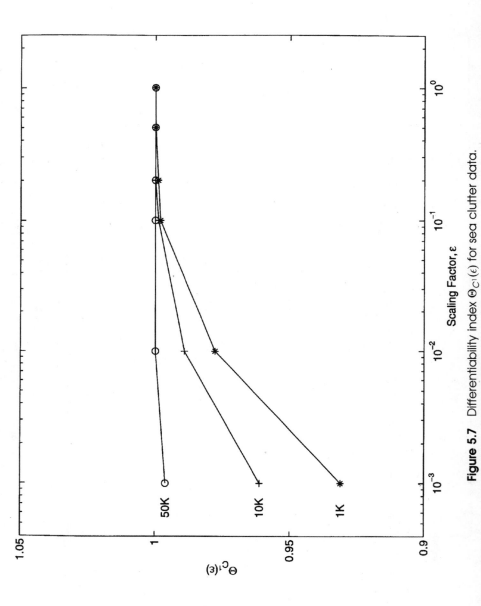

Figure 5.7 Differentiability index $\Theta_{C^1}(\epsilon)$ for sea clutter data.

6

TESTS OF STATIONARITY
AND NONLINEARITY

6.1 INTRODUCTION

As explained previously, a chaotic process is a nonlinear deterministic dynamical process. It follows therefore that when we are given an experimental time series, one of the first tests that should be performed on the time series is that of nonlinearity. It is only when the nonlinear nature of the time series is confirmed that we are in a position to conduct further tests for chaotic evidence.

Another important preliminary test that should be conducted on the time series is that of stationarity. This test is mandatory as in fact all the algorithms currently available for estimating the chaotic invariants of a time series operate on the premise that the time series is stationary. Moreover, the experimental time series must be long enough for the estimation of the chaotic invariants to be reliable.

Before proceeding with the extraction of chaotic invariants from an experimental time series, it is therefore advisable to test for the stationarity and nonlinearity of the time series, as described next.

6.2 TEST OF STATIONARITY

There has been a great deal of interest in nonstationary time series analysis[1] for two reasons. First, many naturally occurring time series are nonstationary. For example, physiological systems tend to vary over multiple time scales. Second, time series from spatially extended physical systems have a tendency to exhibit dynamics over multiple time scales [Bak et al., 1987]. Thus nonstationary time series analysis is of wide applicability. Notwithstanding these physical realities, the fundamental assumption underlying almost all existing linear and nonlinear techniques of time series analysis is that the time series is stationary [Eckmann and Ruelle, 1985; Tong, 1990]. Misleading results can be obtained by applying these techniques to nonstationary time series. Thus the development of techniques for testing the stationarity of the time series is an important preliminary step in time series analysis.

6.2.1 Recurrence Plots

The dynamical analysis of an experimental time series requires it to be drawn from a stationary process. To satisfy this requirement, we may use the so-called recurrence plot due to Eckmann and Ruelle [1987]. This is a graphical tool for the diagnosis of *drift and hidden periodicities* in the time evolution of dynamical systems, which are unnoticeable otherwise. A brief description of the construction of recurrence plots is given in what follows.

Let $s(i)$ be the ith point on the orbit describing a dynamical system in d_E-dimensional space. The recurrence plot is an array of dots in an $N \times N$ square, where a dot is placed at (i, j) whenever $s(j)$ is sufficiently close to $s(i)$. To obtain a recurrence plot from a time series $s(n)$, in practice, we may proceed as follows. First, choosing an embedding dimension d_E, we construct the d_E-dimensional orbit of $s(i)$ by the method of delays (described later). Next, we choose $r(i)$ such that the ball of radius $r(i)$ centered at $s(i)$ in $\mathbf{R}^{d_E}$ contains a reasonable number of other points $s(j)$ of the orbit. Finally, we plot a dot at each point (i, j) for which $s(j)$ is in the

---

[1] From a theoretical point of view a time series $x_1, x_2, \ldots$ is said to be nonstationary if, for some m, the joint probability distribution of $x_i, x_{i+1}, \ldots, x_{i+m-1}$ is dependent on the time index i [Priestley, 1988]. From a practical point of view a time series $x_1, x_2, \ldots, x_N$ is nonstationary if, for low m, there are variations in the estimated joint distributions of $x_i, x_{i+1}, \ldots, x_{i+m-1}$ that occur on time scales of order N.

ball of radius $r(i)$ centered at $\mathbf{s}(i)$. This picture is called a *recurrence plot.* Note that the i and j are measured in units of time. These plots tend to be fairly symmetric with respect to the diagonal $i = j$, because if $\mathbf{s}(i)$ is close to $\mathbf{s}(j)$, then $\mathbf{s}(j)$ is close to $\mathbf{s}(i)$. We cannot expect to get complete symmetry of the pattern because of the fact that $r(i)$ and $r(j)$ may not be equal.

The power and efficacy of this method in extracting the *drift and hidden periodicities* of time series are illustrated here by plotting the recurrence plots of some computer-generated time series. Figure 6.1 shows the recurrence plot for a computer-generated sinusoid (20 Hz, sampled at 256 Hz). The time series is shown in Fig. 6.1a and its recurrence plot in Fig. 6.1b. We used embedding dimension $d_E = 2$, number of nearest neighbors equal to 20, time delay $T = 5$, and number of samples $N = 500$. It may be observed that the large-scale diagonal line segments parallel to $i = j$ indicate the periodic behavior of the data. Figure 6.2b illustrates the recurrence plot for a signal generated by adding 0–80% of the original sinusoidal signal and hence deliberately making it a nonstationary signal. Here, again, we used the parameters as in the above case. The drift in the signal is clearly recognized by the overall reduction of recurrences away from the main diagonal. The fading of the pattern is due to the nonstationarity inherent in the signal. The time series is shown in Fig. 6.2a, which clearly shows the nonstationarity introduced in the signal.

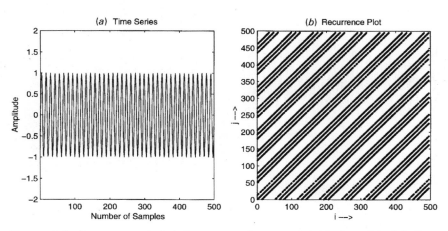

Figure 6.1 Recurrence plot for computer-generated sinusoid: (a) time series; (b) recurrence plot.

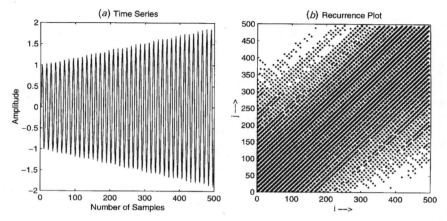

Figure 6.2 Recurrence plot for computer-generated sinusoid plus transient: (*a*) time series; (*b*) recurrence plot.

The third example shown in Fig. 6.3 is due to a white Gaussian random[2] (WGN) sequence actually recorded from a physical device. The uniform distribution of the pattern (see Fig. 6.3*b*) in this case clearly illustrates that it is indeed a stationary process without any periodicity. Here, again, we have chosen the same parameters as those in the previous cases.

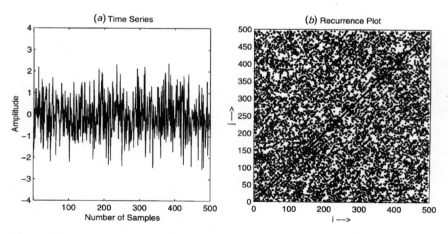

Figure 6.3 Recurrence plot for computer-generated white Gaussian noise: (*a*) time series; (*b*) recurrence plot.

[2] Details of the white-noise-generating device will be given in the next footnote.

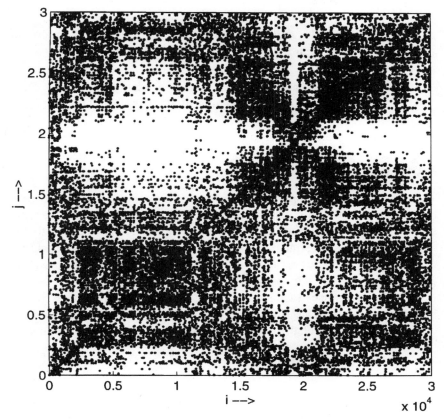

Figure 6.4 Recurrence plot for typical sea clutter data set.

Figure 6.4 shows the recurrence plot for a typical sea clutter data set. Here, we used 30,000 data points with embedding dimension $d_E = 5$, normalized embedding delay $\tau = 11$, and number of nearest neighbors equal to 10. The overall pattern is fairly uniform, confirming the stationarity of the data set.

6.3 TESTS OF NONLINEARITY

As mentioned previously, nonlinearity is one of the essential requirements for a process to be chaotic. We may check for the nonlinearity of an experimental time series in one of three different ways:

1. Using a wave-shaping filter (WSF)
2. Using surrogate data analysis, with the growth of interpoint distances as the discriminating statistic [Schouten et al., 1994] (SIPD)
3. Using surrogate data analysis, with the correlation dimension D_2 as the discriminating statistic [Theiler et al., 1992] (SCD)

These methods are described in turn.

6.3.1 WSF Method

This first method perhaps gives the most convincing proof that a given time series is indeed a nonlinear process. First, a WSF is used to generate a linear process with identically the same magnitude spectrum as that of the time series, as illustrated in Fig. 6.5. Let $S_x(\omega)$ be the power spectrum of the given time series $x(n)$. A linear-phase finite impulse response (FIR) filter is designed in such a way that, when it is stimulated by a zero-mean white Gaussian noise[3] (a linear process), it produces an output signal (a linear process) that has the same magnitude spectrum as that of the original time series. That is, the filter's magnitude response is defined by

$$H(\omega) = \sqrt{\frac{S_x(\omega)}{S_w(\omega)}} \tag{6.1}$$

where $S_w(\omega)$ is the power spectrum of the white Gaussian noise. The filter coefficients are obtained by computing the inverse Fourier transform of $H(\omega)$, assuming a linear phase.

These two signals (i.e., the given time series and the output of the WSF) are then subjected to tests described in Chapter 4. Those tests should reveal that the output of the WSF is drawn from a random process (a known fact) and the given time series is different.

---

[3] For the filter output to be truly stochastic, the stimulus must be stochastic too. This would therefore rule out the use of a pseudo noise sequence generated on a computer, which, in reality, is a deterministic process. To satisfy the requirement that the stimulus in Fig. 6.5 be stochastic, the pure white Gaussian noise was generated using a commercially available analog noise generator (NC 1107A-1, Noise Com Inc.) coupled with an amplifier and A/D converter. This physical device contains a hermetically packaged noise diode that has been burned in for 168 h and operates in a temperature range of -35 to $+100°C$. It produces white Gaussian noise at $+13$ dBm and in a frequency range between 100 Hz and 100 MHz.

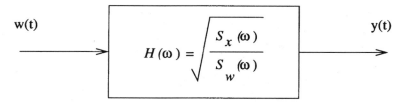

Figure 6.5 Block diagram of wave-shaping filter.

6.3.2 SIPD Method

In this second method [Schouten et al., 1994], we perform a statistical test in which the given time series is compared with surrogate data sets. The surrogate data are created using a stochastic linear model with the same autocorrelation coefficients as those of the given time series. The (exponential) growth of interpoint distances is used as the discriminating statistic to test the null hypothesis that the original time series can be described by linearly correlated noise. A quantity Z is calculated from the Mann–Whitney rank-sum statistic. It is normally distributed with zero mean and unit variance under the null hypothesis that two observed samples of interpoint distances in the original and surrogate data sets come from the same population. A value of Z less than -3.0 is considered to be a valid reason for strong rejection [Siegel, 1956].

6.3.3 SCD Method

This third and final method involves applying the surrogate data analysis proposed by Theiler et al. [1992]. Surrogate data sets are generated and then compared with the original data set using the correlation dimension D_2 as the discriminating statistic. First, the Fourier spectrum of the original data set is computed for positive and negative frequencies (no windowing). This spectrum has a complex amplitude at each frequency. Next, the phase is randomized by multiplying each complex amplitude by $e^{j\phi}$, where ϕ is picked from a uniform distribution in the interval $[0, 2\pi]$. In order for the inverse Fourier transform to be real, the phase is symmetrized by having $\phi(f) = \phi(-f)$. Finally, the inverse Fourier transform of the phase-randomized signal is computed to produce the surrogate data set. The results of D_2 analysis of the original time series and the corresponding surrogate data set should reveal that they are statistically distinct, thus again rejecting the null hypothesis that the original time series is produced by a linear stochastic process.

The step-by-step procedure for generating surrogate data sets from the original data set is listed below:

1. Input the original data into an array $y[t]$, $t = 1, 2, \ldots, N$.
2. Compute the discrete Fourier transform: $z[t] = \text{DFT}(y[t])$.
3. Note $z[t]$ has both real and imaginary components.
4. Randomize the phases: $z'[t] = z[t]e^{i\phi[t]}$.
5. Symmetrize the phases:
   a. $\text{Re } z''[t] = \text{Re}(z'[t] + z'[N + 1 - t])/2$;
   b. $\text{Im } z''[t] = \text{Im}(z'[t] - z'[N + 1 - t])/2$.
6. Invert the DFT: $y'[t] = \text{DFT}^{-1}(z''[t])$.
7. Note that because of the symmetry of the phases, the resulting time series $y'[t]$ is real; the time series so generated is the surrogate data.
8. Repeat the above procedure for generating different sets of surrogate data.

6.4   RESULTS OF NONLINEARITY TESTS

Next, we discuss some of the results we obtained from the nonlinearity tests on clutter data sets. As we mentioned previously, the nonlinearity tests were performed in three different ways.

In the first method, we designed a WSF to filter out a pure Gaussian white sequence (linear stochastic process generated by a commercially available physical device) to produce a signal that has the same power spectrum as that of the clutter data set. We used a data length of 4096 points of the original sea clutter data for designing the WSF filter. Figure 6.6 shows the experimental data details. Figure 6.6a refers to the original sea clutter data, and its power spectrum is shown in Fig. 6.6b. Figure 6.6e is the time series obtained using the WSF stimulated by white Gaussian noise (Fig. 6.6c). Its spectrum is shown in Fig. 6.6f, which is almost the same as that in Fig. 6.6b referring to the original sea clutter data set.

The set of criteria listed in Chapter 4 is applied to the signals shown in Fig. 6.6a (original sea clutter data) and Fig. 6.6e (filtered signal using the WSF), independently, to determine the nature of the underlying physical process by which the signals are being generated. The filtered signal behaved like a noise process, whereas the sea clutter data revealed that it is

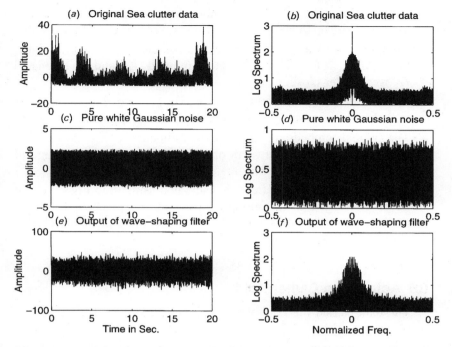

Figure 6.6 Wave-shaping filter procedure: (*a*) I–Q corrected amplitude component of sea clutter; (*b*) corresponding FFT spectrum; (*c*) pure white Gaussian noise; (*d*) corresponding spectrum; (*e*) output of WSF; and (*f*) corresponding spectrum.

indeed a chaotic process. The results of these studies on these two signals are listed below:

(a) Filtered Signal

1. The nonlinearity test using the SIPD algorithm showed that the filtered signal is a linear process (known fact) with a Mann–Whitney rank-sum statistic [Siegel, 1956] (*Z*-value) always positive (>0).
2. The correlation dimension D_2 was fractal but it kept on increasing for increasing embedding dimension (d_E) values. This is again a characteristic of a noise process.
3. The Lyapunov spectrum consisted of several positive exponents and the number of positive exponents kept on increasing for increasing

values of d_E. The sum of all Lyapunov exponents for $d_E < 7$ was always positive, showing the nondissipative nature of the system.

4. Computation of the Kaplan–Yorke dimension (D_{KY}) was not feasible for $d_E < 7$. For higher d_E's, the computed D_{KY} and the estimated D_2 were never close.

5. The estimated Kolmogorov entropy is high, and this value never tallied with the sum of all positive Lyapunov exponents.

6. The global embedding dimension estimated using the GFNN method never converged to an optimum value as the percentage of the FNN was always much greater than zero.

7. The effective degrees of freedom estimated using the LFNN method never converged to an optimum value as the percentage of bad predictions was almost flat for all the embedding dimensions chosen (from 1 to 12).

(b) Actual Sea Clutter Data

1. The data passed the nonlinearity test with a Z-value [Siegel, 1956] always less than -3.0.

2. The D_2 value was fractal, and it saturated to a value between 4.1 and 4.5 for increasing embedding dimension.

3. The Lyapunov spectrum consisted of two positive exponents followed by one exponent close to zero and two or more negative exponents corresponding to embedding dimension 5 or more (the number of exponents is equal to the embedding dimension). The sum of all Lyapunov exponents was always negative, indicating that it is indeed a dissipative system.

4. The computed D_{KY} was always close to the estimated D_2.

5. The estimated Kolmogorov entropy is finite, and its value is close to the sum of all positive Lyapunov exponents.

6. The estimated global dimension converged to 5.

7. The local dimensionality estimated is also 5 and satisfies the condition $d_L \leq d_E$.

The results of this first method show that sea clutter is a nonlinear process.

Next, we applied the second (SIPD) method to test the nonlinearity of the sea clutter data sets [Schouten et al., 1994]. The Mann–Whitney rank-sum statistics (Z-value) for 50 surrogate data sets were less than -3.0 for almost all the data sets used [Siegel, 1956]. Therefore, the test rejected the null hypothesis that each clutter sample is that of a linear process.

In the third (SCD) method, the surrogate data analysis proposed by Theiler et al. [1992] was applied to a minimum of 10 sea clutter data sets. Twenty such phase-shuffled surrogate sets were generated for each clutter data set. The surrogate D_2 (maximum-likelihood estimate) values were statistically compared (Student's t test) with the D_2 values of the original clutter data [Siegel, 1956]. The surrogate sets were significantly different from the original data sets ($p < 0.0001$). Figure 6.7 shows the original data (Fig. 6.7a) and their two surrogates (Figs. 6.7c and 6.7e) along with their Fourier spectra (Figs. 6.7b, d, and f, respectively). The waveform of the surrogate data looks entirely different from that of the original sea clutter signal, but their spectra are essentially identical. The spurious low-

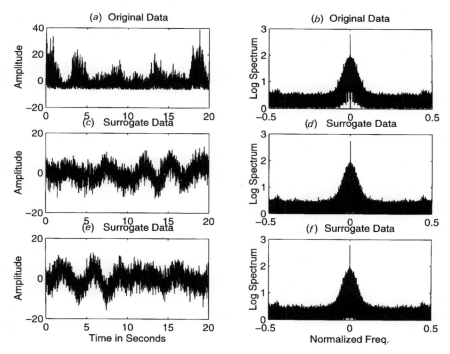

Figure 6.7 Surrogate data analysis: (a) I–Q corrected amplitude component of sea clutter; (b) corresponding FFT spectrum; (c) one set of surrogate data; (d) corresponding spectrum; (e) another set of surrogate data; and (f) corresponding spectrum.

frequency effects observed in Figs. 6.7b, d, and f are due to the limitation of the algorithm [Theiler et al., 1992]. From this nonlinearity test, also, the inference is that sea clutter is a nonlinear process.

Thus these three nonlinearity tests, in their own individual ways, confirm the known fact that the underlying physics of sea clutter is indeed nonlinear. Having satisfied ourselves that the given experimental time series is indeed drawn from a process that is both stationary and nonlinear, we proceed with the important task of extracting chaotic invariants from it.

7

SOME FACTS ABOUT CHAOTIC PROCESSES

7.1 INTRODUCTION

The great power of science lies in the ability to relate cause and effects in physical phenomena. On the basis of laws of gravitation, for example, eclipses can be predicted thousands of years in advance. There are other natural phenomena that are not as predictable. Although the movements of the atmosphere obey the laws of physics just as much as the movements of the planets do, weather forecasts are still stated in terms of probabilities. The weather, the flow of mountain stream, the roll of the dice all have unpredictable aspects. Since there is no clear relation between cause and effect, such phenomena are said to have random elements. Yet, until recently, there was little reason to doubt that precise predictability could in principle be achieved. It was assumed that it was only necessary to gather and process a sufficient amount of information.

Such a viewpoint has been altered by a striking discovery: Simple deterministic systems with only few elements can generate random behavior [May, 1976, and references there in]. The randomness is fundamental, and gathering more information does not make it go away. Randomness generated in this way has come to be called chaos.

There is no accepted general definition of chaos. For an applied scientist, a system is chaotic when it exhibits sensitive dependence on initial conditions; that is, in any neighborhood of a state there exist states whose trajectory diverges in time from that of the initial one. This property leads to "unpredictability" of the evolution of the system. For mathematicians, a system is chaotic if it has a compact invariant set A of states on which the dynamics exhibit sensitive dependence on initial conditions. There is at least a dense orbit in A and the set of periodic orbits is dense in A.

So, roughly speaking, chaos is effectively the unpredictable long-time behavior arising in a deterministic dynamical system because of sensitivity to initial conditions. For a dynamical system to be chaotic, it must have a "large" set of initial conditions that are highly unstable. No matter how precisely we measure the initial condition in these systems, the prediction of its subsequent motion goes radically wrong after a short time. Typically, the predictability horizon grows only logarithmically with the precision of measurement.

More precisely, a map **M** is chaotic on a compact invariant set S if:

- **M** is transitive on S, and
- **M** exhibits sensitive dependence on S.

Transitivity means that if, for a topological space S and for all $u,\ v,\ w \in S$, we have $u \sim v$ and $v \sim w$, then $u \sim w$.

As a consequence of long-term unpredictability, a time series from chaotic systems may appear irregular and disordered. However, chaos is definitely not complete disorder; it is disorder in a deterministic dynamical system, which is always predictable in the short-term sense.

The discovery of chaos has created a new paradigm in scientific modeling. On the one hand, it implies new fundamental limits on the ability to make predictions. On the other hand, the determinism inherent in chaos implies that many random phenomena are more predictable than had been thought. Random-looking information gathered in the past and shelved because it was assumed to be too complicated can now be explained in terms of simple laws. The result is a revolution that is affecting many different branches of science.

7.2 SENSITIVE DEPENDENCE ON INITIAL CONDITIONS[1]

A seeming paradox is that chaos is deterministic, generated by fixed rules that do not subject themselves to any element of chance. In principle, the future is completely determined by the past, but in practice, small uncertainties are amplified, so that even though the behavior is predictable in the short term, it is unpredictable in the long term. This distinct phenomenon is referred to as the *sensitive dependence on initial conditions*.

We present here the outlook of two luminaries, Laplace (eighteenth century) and Poincaré (twentieth century) on chance and probability. The eighteenth-century mathematician Pierre Simon de Laplace once boasted that, given the position and velocity of every particle in the universe, he could predict the future for the rest of time. The quotation of Laplace [Crutchfield et al., 1986] is presented here below:

> The present state of the system of nature is evidently a consequence of what it was in the preceding moment, and if we conceive of an intelligence which at a given instant comprehends all the relations of the entities of this universe, it could state the respective positions, motions and general effects of all these entities at any time in the past or future.

Laplace claimed that the laws of nature imply strict determinism and complete predictability, although imperfections in observations make the introduction of probabilistic theory necessary. For more than 100 years his theory ruled, and in the twentieth century, science saw the downfall of Laplacian determinism for two very different reasons.

The first reason is quantum mechanics. The central dogma of this theory is the Heisenberg uncertainty principle, which states that there is a fundamental limitation to the accuracy with which the position and velocity of a particle can be measured. Such uncertainty gives a good explanation for some random phenomena, such as radioactive decay. A nucleus is so small that the uncertainty principle puts a fundamental limit on the knowledge of its motion, and so it is impossible to gather enough information to predict when it will disintegrate.

The source of unpredictability on a large scale must be sought elsewhere, however. Some large-scale phenomena are predictable and others

---

[1] Much of the material in this section follows Crutchfield et al., 1986.

are not. The distinction has nothing to do with quantum mechanics. The trajectory of a cricket ball, for example, is inherently predictable. The trajectory of a flying balloon with the air rushing out of it, in contrast, is not predictable. The balloon obeys Newton's laws just as much as the cricket ball does; then why is its behavior so much harder to predict than that of the ball?

The classic example of such a dichotomy is fluid motion. Under some circumstances, the motion of the fluid is laminar—even, steady, and regular and easily predicted from equations. Under other circumstances, fluid motion is turbulent—uneven, unsteady, and irregular and difficult to predict. The transition from laminar to turbulent behavior is a familiar phenomenon. What causes the essential difference between laminar and turbulent motion?

The late Soviet physicist Lev D. Landau is credited with an explanation of random fluid motion. He explains it in the following way: The motion of a turbulent fluid contains many different, independent oscillations. As the fluid is made to move faster, causing it to become more turbulent, the oscillations enter the motion one at a time. Although each oscillation may be simple, the complicated combined motion renders the flow impossible to predict.

Landau's theory was disproved in the early twentieth century when a French mathematician, Henri Poincaré, realized that random behavior occurs even in very simple systems, without any need for complication or indeterminacy. He noted that unpredictable, "fortuitous" phenomena may occur in systems where a small change in the present causes a much larger change in the future. His 1903 quotation [Crutchfield et al., 1986] is presented below:

A very small cause which escapes our notice determines a considerable effect that we cannot fail to see, and then we say that the effect is due to chance. If we knew exactly the laws of nature and the situation of the universe at the initial moment, we could predict exactly the situation of that same universe at a succeeding moment. But even if it were the case that the natural laws had no longer any secret for us, we could still only know the initial situation *approximately*. If that enabled us to predict the succeeding situation with the *same approximation*, that is all we require, and we should say that the phenomenon had been predicted, that is governed by laws. But it is not always so; it may happen that small differences in the initial conditions produce very great ones in the final phenomena. A small error in the former will produce an enormous error in the latter. Prediction becomes impossible, and we have the fortuitous phenomenon.

This quotation of Poincaré foreshadows the contemporary view that arbitrary small uncertainties in the state of a system may be amplified in time, and so predictions of the distant future cannot be made.

The notion is clear if you consider the example explained here. A boulder is precariously perched on the top of an ideal hill. A slightest push will cause the boulder to roll down on either side of the hill. The direction of the boulder depends sensitively on the direction of the push and the push can be arbitrarily small. This means that a tiny difference in the input (here, the direction and strength of push) may be quickly amplified to have overwhelming difference in the output. This is the so-called butterfly effect.

Mathematically, a set S exhibits sensitive dependence on initial conditions if there is an r such that, for any $\varepsilon > 0$ and for each $x \in S$, there is a y such that $|x - y| < \varepsilon$ and $|x - y| > r$ for some number of steps $n > 0$. This means that there is a fixed distance r such that no matter how precisely we specify an initial state there are nearby states that eventually get a distance r away.

It is this exponential amplification of errors due to chaotic dynamics that provides the second reason for Laplace's undoing. Quantum mechanics implies that initial measurements are always uncertain, and chaos ensures that the uncertainties will quickly overwhelm the ability to make predictions. Without chaos Laplace might have hoped that errors would remain bounded, or at least grow slowly enough to allow him to make predictions over a long period. With chaos, predictions are rapidly doomed to gross inaccuracy in the long term.

We shall illustrate this phenomenon of sensitive dependence on initial conditions with the use of a simple nonlinear equation: the logistic equation from mathematical biology. The logistic equation is a model for the growth of an idealized population consisting of only one species. This system was put forth as a simple model for population growth by the ecologist Robert May [1976]. To be precise, let us suppose that there is a single species whose population grows and dwindles over time in a controlled environment. Suppose we measure the population of the species at the end of each generation. Rather than produce the actual count of individuals present in the colony, suppose we measure instead the percentage of some limiting number or maximum population. That is, let us write A_n for the fraction of population after generation n, where $0 \leq A_n \leq 1$. One simple rule that an ecologist may use to model the growth of this population is the *logistic equation*

$$A_{n+1} = kA_n(1 - A_n) \tag{7.1}$$

where k is some constant that depends on ecological conditions such as the amount of food present. Using this formula, the population in the succeeding generation may be deduced from a knowledge of only the population in the preceding generation and the constant k.

Note how trivial this formula is. It is a simple quadratic formula in the variable A_n. Given A_n and k, A_{n+1} can be computed exactly. Table 7.1 shows several such predicted values for different values of k. We can see several things happen here. When k is small, the fate of the population seems quite predictable. Indeed, for $k = 0.5$, the population dies out. For $k = 1.2$, 2, 3, it tends to stabilize or reach a definite limiting value. For values of $k > 3$, different values of k give startlingly different results. For $k = 3.1$, the limiting values tend to oscillate between two values. For $k = 3.4$, the limiting values tend to oscillate between four distinct values. Finally for $k = 4$, there is no apparent pattern to be discerned. One initial value, $A_0 = 0.5$, leads to the disappearance of the species after only two

Table 7.1   Values of A_n for Various k-Values

$k = 0.5$	$k = 1.2$	$k = 2.0$	$k = 3.0$	$k = 3.1$	$k = 3.4$	$k = 4.0$	$k = 4.0$
0.500	0.500	0.500	0.500	0.500	0.500	0.400	0.500
0.125	0.300	0.500	0.675	0.775	0.850	0960	1.000
0.055	0.252	0.500	0.592	0.540	0.434	0.154	0.000
0.026	0.226	0.500	0.652	0.770	0.835	0.520	0.000
0.013	0.210	0.500	0.613	0.549	0.469	0.998	0.000
0.006	0.199	0.500	0.641	0.768	0.847	0.006	0.000
0.003	0.191	0.500	0.622	0.553	0.441	0.025	0.000
0.002	0.186	0.500	0.635	0.766	0.838	0.099	0.000
0.001	0.181	0.500	0.626	0.555	0.461	0.358	0.000
0.000	0.178	0.500	0.632	0.766	0.845	0.919	0.000
0.000	0.176	0.500	0.628	0.556	0.446	0.298	0.000
0.000	0.174	0.500	0.631	0.765	0.840	0.837	0.000
0.000	0.172	0.500	0.629	0.557	0.457	0.547	0.000
0.000	0.171	0.500	0.630	0.765	0.844	0.991	0.000
0.000	0.170	0.500	0.629	0.557	0.448	0.035	0.000
0.000	0.170	0.500	0.630	0.765	0.841	0.135	0.000
0.000	0.169	0.500	0.629	0.557	0.455	0.466	0.000
0.000	0.168	0.500	0.630	0.765	0.843	0.996	0.000
0.000	0.168	0.500	0.629	0.557	0.450	0.018	0.000
0.000	0.168	0.500	0.630	0.765	0.851	0.070	0.000
0.000	0.168	0.500	0.630	0.557	0.455	0.261	0.000
0.000	0.168	0.500	0.630	0.765	0.843	0.773	0.000

generations, whereas $A_0 = 0.4$ leads to a population count that seems to be completely random.

This is the unpredictable nature of the logistic process with respect to slightly different initial conditions. We show this process as a picture in Fig. 7.1, where the x axis is the values of k and the y axis is the predicted values of the population for a starting value, say $A_0 = 0.4$.

This example illustrates one of the major consequences of the discovery of chaos in deterministic systems:

- Small changes in the initial conditions may lead to vastly different eventual values for the population.

The existence of chaos in deterministic systems has a number of important consequences in both mathematics and the physical sciences. First, it means that no matter how accurately we measure the physical quantities that determine the system, we may never be able to accurately predict the resulting motion. Second, it indicates that the search for individual, specific solutions to the system may be useless. After all, any small change will produce a vastly different solution, perhaps necessitating completely new methods of analysis in order to generate the solution. This may be impractical or may be even impossible.

What does a scientist do in the face of chaos? Obviously, by the very nature of a chaotic system, the search for a specific solution of the equations is not especially fruitful. Hence the scientist takes a more global viewpoint. Instead of seeking specific solutions, the scientist seeks to describe the totality of all possible solutions. Although the particular behavior of a solution may be unpredictable, the totality of all these solutions may be identifiable.

As an example, the meteorologist may not be able to predict whether it will be rainy or sunny on a given day in August in New York, but he knows that it will not be snowing on that day. Thus, there are limits to the unpredictability of a system, and finding these limits is an important task. Toward that end, there have been a number of remarkable advances in recent years.

7.3 NONLINEAR DYNAMICAL SYSTEMS

Recent advances in the branch of mathematics and physics known as *nonlinear dynamical systems* promise to revolutionize the way scientists

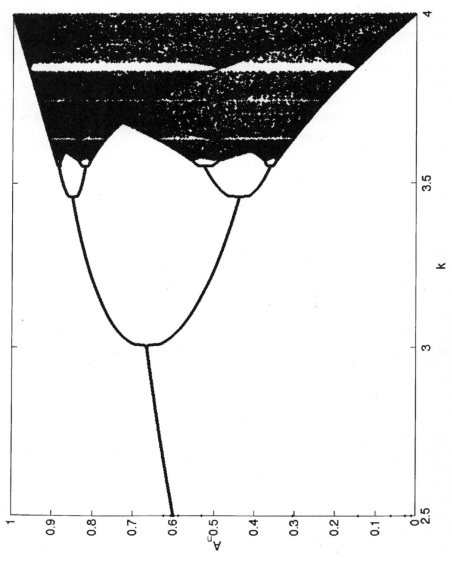

Figure 7.1 Logistic map showing routes to chaos as successive bifurcations.

view many different kinds of evolutionary processes. These processes occur in all branches of science, ranging from fluctuations of temperature, pressure, wind speed, and so on in meteorology to the ups and downs of the stock market in economics. Any physical, chemical, or biological process that evolves in time is an example of a dynamical system. Some systems are simple, like the motion of a pendulum, which gradually damps down to a stable resting position. Others are much more complicated, like the motion of galaxies in the universe or a fluid tumbling over an obstacle.

A dynamical system consists of an abstract *phase space* or *state space*, whose coordinates describe the dynamical state at any instant of time, and a dynamical rule that specifies the immediate future trend of all state variables given only the present values of those state variables. Mathematically, a dynamical system is described by an initial-value problem.

Dynamical systems are "deterministic" if there is a unique consequent to every state and "stochastic" or "random" if there is more than one consequent chosen from some probability distribution. A dynamical system can be continuous or discrete in time. A continuous-time dynamical system in an N-dimensional phase space can be represented by a set of N first-order, autonomous, ordinary differential equations. Using vector notation, this can be written in the following way:

$$\frac{d\mathbf{x}(t)}{dt} = \mathbf{F}[\mathbf{x}(t)] \tag{7.2}$$

where $\mathbf{x}$ is an N-dimensional vector. The term $\mathbf{F}[\mathbf{x}]$ is a "vector field" and gives a vector pointing in the direction of the velocity at every point in phase space. Given any initial state of the system, in principle, any future state of the system can be derived using the above equation.

The path generated by the system through its evolution in phase space is referred to as the "orbit" or "trajectory." It is also common to refer to the time dynamical system as a "flow."

A discrete-time dynamical system is commonly referred to as a "map." Here, the time index is integer valued rather than continuous. Such a system can be described as follows:

$$\mathbf{x}_{n+1} = \mathbf{M}(\mathbf{x}_n) \tag{7.3}$$

where $\mathbf{x}_n$ has N components. Here, again, any future state of the system can be derived from the above description, given any initial state of the system.

It is the basic goal of scientists who work with dynamical systems to develop methods that accurately predict the future behavior of the system. Sometimes this may be accomplished by direct observation or experimentation: Years of waking up each morning have convinced us that the sun will rise again the next day. But other evolutionary processes like the weather or the stock market are much more difficult to predict. So, scientists seek other methods to make predictions.

One of the most common methods for making predictions is first to set up a mathematical model of the system at hand and then to solve the resulting equations. The scientist uses accepted laws such as Newton's laws or Hooke's law to set up this mathematical model, which is often a differential equation or a difference equation as discussed above. The solutions of these equations then yield the desired predictions. Sometimes the solutions of these equations are relatively easy, as could be the case when the equations are linear. When the equations are nonlinear, the situation changes drastically. There are very few mathematical techniques available for solving such equations explicitly. Thus, the scientists must use other methods to "solve" the equations. Very often this necessitates the use of a computer and various approximation techniques to gain at least partial insight into the solution. But this approach, as we shall see, is not altogether satisfactory. One of the main advances in recent years in the study of nonlinear dynamics is the recognition that such computer solutions may be totally meaningless. Many systems behave so erratically or unpredictably that the slightest error or approximation used in their formulation or solution leads to completely erroneous predictions. This phenomenon is nothing but the sensitive dependence on initial conditions described in the previous section, which is one of the basic characteristics of chaotic processes.

7.4   WHAT IS PHASE SPACE?

A state space is a useful concept for visualizing the behavior of a dynamical system. It is the collection of all possible states of a dynamical system. This can be finite (e.g., state variables of coin tossing), countably infinite (e.g., state variables are integers), or uncountably infinite (e.g., state variables are real numbers). Implicit in the notion is that a particular state in phase space specifies the system completely. The usefulness of the state-space picture lies in the ability to represent behavior in a geometric form. The motion of a pendulum, for example, is completely determined by its initial position and velocity.

7.5   WHY A CHAOTIC APPROACH?

Chaos is the irregular behavior of simple equations, and irregular behavior is ubiquitous in nature. One of the primary motivations for studying chaos is this: Given an observation of irregular behavior, is there a simple explanation that can account for it? And if so, how simple is it? There is a growing consensus that a useful understanding of the physical world will require more than finally uncovering the fundamental laws of physics. Simple systems, which obey simple laws, can nonetheless exhibit exotic and unexpected behavior. Nature is full of surprises that turn out to be direct consequences of Newton's laws.

Daily experience shows that for many physical systems small changes in the initial conditions lead to small changes in the outcome. But there are cases for which the opposite of this rule is true. It has become clear in recent years, partly triggered by the studies of nonlinear systems using supercomputers, that a sensitive dependence on the initial conditions that results in a chaotic time behavior is by no means exceptional but a typical property of many systems. Such behavior, for example, has been found in periodically simulated cardiac cells, in electronic circuits, at the onset of turbulence in fluids and gases, in chemical reactions, in lasers, and so on. Mathematically, all nonlinear dynamical systems with more than two degrees of freedom, that is, especially many biological, meteorological, or economic models, can display chaos and therefore become unpredictable over longer time scales.

Deterministic chaos is now a very active field of research with many exciting results. Methods have been developed to classify different types of chaos, and it has been discovered that many systems show, as a function of an external control parameter, similar transitions from order to chaos. This universal behavior is reminiscent of ordinary second-order phase transitions, and the introduction of renormalization and scaling methods from statistical mechanics has brought new perspectives into the study of deterministic chaos.

During the past few decades, a growing number of systems have been shown to exhibit randomness due to simple chaotic attractor. It should be emphasized, however, that chaos theory is far from a panacea. Many degrees of freedom can also give rise to complicated motions that are effectively random.

One usually measures the complexity of a physical system by the number of degrees of freedom that the system possesses. However, it is useful to distinguish nominal degrees of freedom from effective degrees of freedom. Although there may be many nominal degrees of freedom available, the physics of the system may organize the motion into only a few effective degrees of freedom. Distinguishing a behavior that is irregular but low dimensional from a behavior that is irregular because of stochasticity (many effective degrees of freedom) is the motivation for a chaotic approach. Estimating dimension from an observed data is one way to detect and quantify the chaotic properties of natural and artificial complex systems.

8

RECONSTRUCTION OF EMBEDDING SPACE

8.1 INTRODUCTION

Reconstruction of the embedding space (phase space) is an important step in the application of dynamical system techniques to experimental data, usually obtained by "measuring" the value of a single observable as a function of time. An important tool in a dynamic toolkit is "delay coordinate embedding," which creates a phase-space portrait from a single time series. It seems remarkable that one can reconstruct a picture equivalent (topologically) to the full Lorenz attractor in three-dimensional phase by measuring only one of its coordinates, say $x(n)$, and plotting the delay coordinates $(x(n), x(n + \tau), x(n + 2\tau))$ for a certain τ.

The reconstruction of an embedding space is the focus of attention for the material presented in this chapter.

8.2 MEASUREMENTS AND STATE REPRESENTATION

In a laboratory, it is seldom the case that all relevant dynamical variables can be measured in an experiment. How can we proceed to study the dynamics in such a situation? A key element in resolving this general class of problems is provided by the embedding theory [Takens, 1981]. In typical situations, points on the dynamical attractor in the full system phase space have a one-to-one correspondence with measurements of a limited number of variables. This is a powerful fact. By definition, a point in the state space carries complete information about the current state of the system. If the equations defining the system dynamics are not explicitly known, this phase space is not directly accessible to the observer. A one-to-one correspondence means that the state space can be identified by measurements.

Assume that we can simultaneously measure m variables $y_1(t), y_2(t), \ldots, y_m(t)$, which we denote by the vector $\mathbf{y}(t)$. This m-dimensional vector can be viewed as a function of the system state $\mathbf{x}(t)$ in the full system phase space:

$$\begin{aligned} \mathbf{y} &= \mathbf{F}(\mathbf{x}) \\ &= (f_1(\mathbf{x}), \ldots, f_m(\mathbf{x})) \end{aligned} \tag{8.1}$$

We call the function $\mathbf{F}$ the *measurement function* and the m-dimensional vector space in which the vector $\mathbf{y}$ lies the *reconstruction space*.

We have grouped the measurements as a vector-valued function $\mathbf{F}$ of $\mathbf{x}$. The fact that $\mathbf{F}$ is a function is a consequence of the definition of state: Information about the system is determined uniquely by the state, so each measurement is a well-defined function of $\mathbf{x}$.

As long as m is taken sufficiently large, the measurement function $\mathbf{F}$ generically defines a one-to-one correspondence between the attractor states in the full state space and m-dimensional vector $\mathbf{y}$. By "one to one" we mean that for a given $\mathbf{y}$ there is a unique $\mathbf{x}$ on the attractor such that $\mathbf{y} = \mathbf{F}(\mathbf{x})$. When there is a one-to-one correspondence, each vector formed of m measurements is a proxy for a single state of the system, and the fact that the entire system information is determined by a state $\mathbf{x}$ is viewed to be true as well as for the measurement vector $\mathbf{F}(\mathbf{x})$. In order for this to be so, it turns out that it is enough to take m larger than twice the box-counting dimension of the attractor.

The one-to-one property is useful because the state of a deterministic dynamical system, thus its future evolution, is completely determined by a point in the full state space. Suppose that when the system is in a given state $\mathbf{x}$ one observes the vector $\mathbf{F}(\mathbf{x})$ in the reconstruction space, and this is followed 1 s later by a particular event. If $\mathbf{F}$ is one to one, each appearance of the measurements represented by $\mathbf{F}(\mathbf{x})$ will be followed 1 s later by the same event. This is because there is one-to-one correspondence between the attractor states in state space and their image vectors in reconstruction space. Thus there is predictive power in measurements **y** being matched to the system state **x** in a one-to-one manner.

8.2.1 Embedding

Two different types of embedding are relevant to attractor reconstruction:

1. *Topological Embedding:* This is nothing more than a one-to-one continuous correspondence between vectors.
2. *Differentiable Embedding:* This means that the differential structure of the attractor is preserved, including quantities such as the Lyapunov exponents.

In what follows, we briefly discuss these two types of embedding.

Topological Embedding

Consider a k-dimensional Euclidean space R^k. Points in this space are specified by giving k real coordinate values, say $x_1, x_2, \ldots, x_k$. Let us represent these elements as a vector **x**. Let **F** be a continuous function from R^k to R^m, where R^m is an m-dimensional Euclidean space. The mapping can be represented in the following way:

$$\mathbf{y} = \mathbf{F}(\mathbf{x}) \tag{8.2}$$

where **y** is in R^m. If A is a subset of R^k, we denote by $\mathbf{F}(A)$ the set of all points $\mathbf{y} = \mathbf{F}(\mathbf{x})$ generated as **x** ranges over A. We call $\mathbf{F}(A)$ the image of A in R^m. The function **F** is one to one on A if given any **y** in the image $\mathbf{F}(A)$ there is one and only one **x** in A such that $\mathbf{y} = \mathbf{F}(\mathbf{x})$. This means that if $\mathbf{x}_1$ and $\mathbf{x}_2$ are both in A, then $\mathbf{F}(\mathbf{x}_1) = \mathbf{F}(\mathbf{x}_2)$ clearly implies that $\mathbf{x}_1 = \mathbf{x}_2$. If **F** is one to one, then the inverse map $\mathbf{F}^{-1}$ can be defined.

In a typical experimental situation, the set A we are interested in is an attractor, which is a compact subset of R^k that is invariant under the dynamical system. The goal is to use measurements to construct the function **F** so that $\mathbf{F}(A)$ is a copy of A that can be analyzed. A finite time series of measurements will produce a finite set of points that make up $\mathbf{F}(A)$. If enough points are present, we may hope to discern some of the properties of $\mathbf{F}(A)$ and therefore that of A.

For a continuous one-to-one map of a compact set, the inverse map $\mathbf{F}^{-1}$ will be continuous as a map on $\mathbf{F}(A)$. A one-to-one map on A that is continuous and has a continuous inverse is called a topological embedding of A. Now, the aim is to find a condition on the set A and the function $\mathbf{F}$ that makes it virtually certain that $\mathbf{F}$ is a topological embedding of A.

For example, consider that A is a finite-length segment of a curve in R^k. Let $\mathbf{F}$ be a function from R^k to R^1, the real line. Depending on $\mathbf{F}$, the set A may or may not be topologically embedded in the real line. Figure 8.1a shows an embedding. In Fig. 8.1b, the function $\mathbf{F}$ fails to embed A. For the latter function, pairs of points that are far apart on A are brought together on $\mathbf{F}(A)$, violating the one-to-one condition. Note also that the function in Fig. 8.1b cannot be easily fixed up to be an embedding. Small perturbations in $\mathbf{F}$ may change the details of the overlap, but it would take a significant overhaul to remove the overlap entirely.

The situation is similar if we consider the image of a curve segment under functions $\mathbf{F}$ from R^k to R^2. Although it is possible to conceive of functions $\mathbf{F}$ that are topological embeddings of the curve segment A, as in Fig. 8.2a, there are still others, exemplified by Fig. 8.2b, that are not embeddings. Furthermore, this cannot be changed to be an embedding by small perturbations. In Fig. 8.2b, the function $\mathbf{F}$ fails to be one to one at the self-intersection point of $\mathbf{F}(A)$, denoted by a dot in the figure.

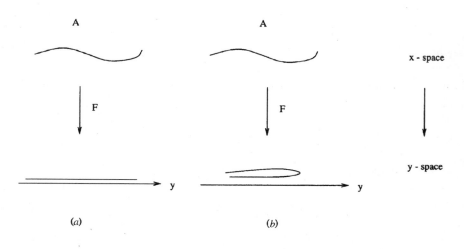

Figure 8.1 Embedding and nonembedding of curve segment: (a) image of A is topological embedding; (b) neither $\mathbf{F}$ nor small perturbations of it are embeddings.

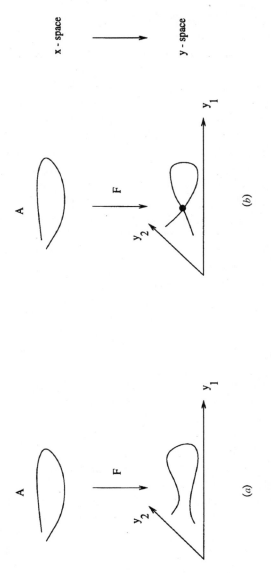

Figure 8.2 Mapping of curve segment to two-dimensional space: (*a*) image of A is topological embedding; (*b*) neither **F** nor small perturbations of it are embeddings.

For functions from R^k to R^3, however, the situation is significantly different: Virtually all functions applied to a curve segment A are topological embeddings. It is still possible to imagine a function **F** that fails to be an embedding, as in Fig. 8.3b. However, in three dimensions, there clearly exists a small perturbation of **F** that pulls the self-intersection of **F**(A) apart at the point P, as in Fig. 8.3a. That is, given any function **F**, either **F** is an embedding of the curve segment A or there exists a small perturbation of **F** that is an embedding.

This is a general property of smooth functions from R^k to R^m. Compact submanifolds of dimension d within R^k will be topologically embedded in R^m as long as $m > 2d$. This fact was proved in the stronger sense of differentiable embedding by Whitney [1936]. In the case of Fig. 8.3, $m = 3$ and $d = 1$. Figure 8.3b illustrates an exceptional case for **F**, in that the image of the curve is not an embedding. But almost every perturbation of **F** (all those that do not cause the self-intersection P to persist) results in a topological embedding.

This fact, that for any given map **F** of a curve segment to three-dimensional space, almost every perturbation is "good" in the sense that it is a topological embedding of A, is obvious from Fig. 8.3. In order to handle cases that are not quite clear-cut (e.g., when A is fractal), it is important to pin down more precisely what we mean by "almost every" in this situation.

A property of functions is called generic in C^n topology[1] if for every function **F**($\mathbf{x}$) that does not have the property there exists a perturbation δ**F**($\mathbf{x}$) such that the magnitude of δ**F**($\mathbf{x}$) and the magnitudes of the derivatives of δ**F**($\mathbf{x}$) up to order n are arbitrarily small and such that **F** $+ \delta$**F** has the property. That is, for any ε, no matter how small, we can find a δ**F** whose Taylor series coefficients of order n or less are all less than ε and such that **F** $+ \delta$**F** has the property.

As D. Ruelle (1989) has written, "It is good to know whether a property is generic or not, but it should be understood that generic does not imply usually true." For example, generic subsets of real numbers can have arbitrarily small probability (Lebesgue measure). Recently, a stronger characterization was introduced, called *prevalence*. A property is called

---

[1] The usage of the term *generic* is equivalent to "dense." Generic is sometimes used in the following more restrictive sense: A property is generic if the set of functions that possess the property is a residual set, which is a set that is a countable intersection of dense open subsets of the function space. In particular, this implies the set is dense, which corresponds to the usage in this book.

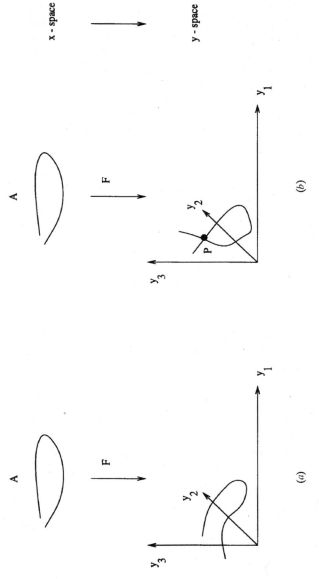

Figure 8.3 Mapping of curve segment into three-dimensional space: (*a*) **F** is an embedding; (*b*) **F** is not an embedding although small perturbations of it yield an embedding.

prevalence among C^n functions if whenever **F** does not have the property, not only do arbitrarily small C^n perturbations have the property, but small perturbations have the property with probability 1. For a precise definition and discussion of "probability 1" in this context, see Sauer et al. [1991]. If a property is prevalent for functions **F**, we often say that almost every function **F** has the property, or that **F** has the property with probability 1.

For functions from R^k to R^m, the property that **F** is a topological embedding of a d-dimensional compact submanifold A is a prevalent (and therefore generic) property of C^1 functions if $m > 2d$. In fact, a very simple set of first-order perturbation suffices. If we consider the finite-dimensional space of $m \times k$ matrices **L**, then no matter what **F** does to the submanifold A, the function $\mathbf{F(x) + Lx}$ is a topological embedding of A for almost every **L** in the sense of probability.

In the discussion so far, we have considered A to be a compact smooth submanifold of the k-dimensional phase space. In many applications, A will be the attractor of a dynamical system and may not be a manifold, or even have an integer dimension. Somewhat surprisingly, if the box-counting dimension was used, the requirement for embedding a fractal set is the same as for manifolds—namely, that the number of measurements m is greater than twice the box-counting dimension of the set A. This leads to the following general statement pertaining to the topological embedding with simultaneous measurements:

> Assume that A is a compact subset of R^k of box-counting dimension D_0. If $m > 2D_0$, then almost every C^1 function $\mathbf{F} = (f_1, f_2, \dots, f_m)$ from R^k to R^m is a topological embedding of A into R^m.

The intuitive reason for the condition $m > 2D_0$ can be seen by considering generic intersections of smooth surfaces in m-dimensional Euclidean space R^m. Two sets of dimensions d_1 and d_2 in R^m may or may not intersect. If they do intersect and the intersection is generic, then they will meet in a surface of dimension

$$d_I = d_1 + d_2 - m \tag{8.3}$$

If this number is negative, generic intersections do not occur. If the surface lies in a special position relative to one another, the intersection may be special and have a different dimension.

Figure 8.4 shows some examples of generic and nongeneric intersections. Consider the example shown in Fig. 8.4a, involving two circles in the plane. If the circles intersect, they will generically meet in a set of dimension $1 + 1 - 2 = 0$ (i.e., a finite set of points), unless the circles lie on top of one another. In the latter nongeneric case, almost every smooth perturbation of the position of the circles results in zero-dimensional intersection. The exceptional cases, which form a probability zero set of perturbations, are those that perturb the circles in lockstep, so that they remain atop one another. For circles in three-dimensional space, since $1 + 1 - 3 < 0$, generic intersections do not occur. Almost every perturbation of intersecting circles in R^3 breaks the intersection. Generic intersections for cases shown in Figs. 8.4b and 8.4c and the nongeneric intersection for the case shown in Fig. 8.4d are self-explanatory.

The requirement $m > 2d$ can now be viewed as the necessary condition for the image $\mathbf{F}(A)$ not to intersect itself. For a d-dimensional set mapped

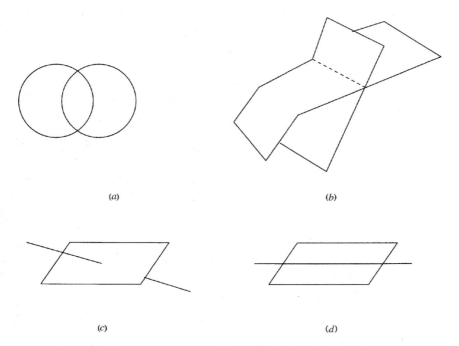

(a)

(b)

(c)

(d)

Figure 8.4 Examples of generic and nongeneric intersections. Generic intersections: (a) $d_1 = d_2 = 1$, $m = 2$; (b) $d_1 = d_2 = 2$, $m = 3$; (c) $d_1 = 2$, $d_2 = 1$, $m = 3$. (d) Nongeneric intersection: $d_1 = 2$, $d_2 = 1$, $m = 1$ but $d_I = 1$.

into R^m, in order to make sure that we can perturb away unlucky intersections of one part of $\mathbf{F}(A)$ with another part of itself, we must require $d + d - m < 0$, or $m > 2d$.

For fractal sets, if d is taken to be the box-counting dimension $d = D_0$, the requirement $m > 2d$ for avoiding self-intersection still holds, although it is harder to draw pictures. This follows from the fact that two generically intersecting fractal sets in R^m of box-counting dimensions d_1 and d_2 will not intersect if $m > d_1 + d_2$.

Delay Coordinates

In the previous section, we described how the phase space of possible states for a dynamical attractor can be reconstructed using m simultaneous measurements that are not in a special position with regard to one another. Although it seems impossible to specify in exact terms what "not in a special position" means in this context, we found that in theory almost every measurement function $\mathbf{F}$ in Eq. (8.1) has this property. In fact, we found that small linear perturbations are sufficient to destroy the special position in any given $\mathbf{F}$, at least in theory. The number of simultaneous measurements required is $m > 2D_0$, where D_0 is the box-counting dimension of the attractor.

Now, assume that our ability to make measurements of independent components is limited. In the worst case, we may be able to make measurements of a single scalar variable, say $y(t)$. Since the measurement depends only on the system state, we can represent such a situation by $y(t) = f(\mathbf{x}(t))$, where f is the single measurement function, evaluated when the system is in state $\mathbf{x}(t)$. We assign to $\mathbf{x}(t)$ the delay coordinate vector

$$\begin{aligned} \mathbf{H}(\mathbf{x}(t)) &= (y(t - \tau), \ldots, y(t - m\tau)) \\ &= (f(\mathbf{x}(t - \tau)), \ldots, f(\mathbf{x}(t - m\tau))) \end{aligned} \tag{8.4}$$

Note that for an invertible dynamical system, given $\mathbf{x}(t)$, we can view the state $\mathbf{x}(t - \tau)$ at any previous time $t - \tau$ as being a function of $\mathbf{x}(t)$. This is true because we can start at $\mathbf{x}(t)$ and use the dynamical system to follow the trajectory backward in time to the instant $t - \tau$. Hence, $\mathbf{x}(t)$ uniquely determines $\mathbf{x}(t - \tau)$. To emphasize this, we define $h_1(\mathbf{x}(t)) = f(\mathbf{x}(t - \tau)), \ldots, h_m(\mathbf{x}(t)) = f(\mathbf{x}(t - m\tau))$. Then, if we write $\mathbf{y}(t)$ for $(y(t - \tau), \ldots, y(t - m\tau))$, we can express Eq. (8.1) as

$$\mathbf{y} = \mathbf{H}(\mathbf{x}) \tag{8.5}$$

where $\mathbf{H}(\mathbf{x}) = (h_1(\mathbf{x}), \ldots, h_m(\mathbf{x}))$.

The delay coordinate function **H** can be viewed as a special choice of the measurement function **F** in Eq. (8.1). For this special choice, the requirement that the measurements do not lie in a special .position is brought into question, since the components of the delay coordinate vector in Eq. (8.4) are constrained: They are simply time-delayed versions of the same measurement function f. This is relevant when considering small perturbations of **H**. Small perturbations in the measuring process are introduced by way of the scalar measurement function f and influence coordinates of the delay coordinate function **H** in an interdependent way. Although it was determined in the simultaneous measurement case that almost every perturbation of h_1, $\ldots$, h_m leads to a one-to-one correspondence, these independent perturbations may not be achievable by perturbing the single measurement function f.

A simple example will illustrate this point. Suppose the set A contains a single periodic orbit whose period is equal to the delay time τ. This turns out to be a bad choice of τ, since each delay coordinate vector from the periodic orbit will have the form $(h(\mathbf{x}), \ldots, h(\mathbf{x}))$ for some $\mathbf{x}$ and lie along the straight line $y_1 = \cdots = y_m$ in R^m. But a circle cannot be continuously mapped to a line without points overlapping, violating the one-to-one property. Notice that this problem will afflict all measurement functions h, so that perturbing h will not help.

In this case, we cannot perturb our way out of the problem by making the measurement function more generic; the problem is built-in. Although this case is a particularly bad one because of a poor choice of the time delay τ, it shows us that the reasoning for the simultaneous measurement case does not extend to delay coordinates, since it gives an obviously wrong conclusion in this case. This problem can be avoided, for example, by perturbing the time delay τ (if indeed that is possible in the experimental settings). In any case, a little extra analysis beyond the geometric arguments we made for the simultaneous measurements case needs to be done. This analysis was begun by Takens [1981] and extended later by Sauer et al. [1991]. The result can be stated as follows:

Topological Embedding: Assume that a continuous-time dynamical system has a compact invariant set A (e.g., A may be a chaotic attractor) of box-counting dimension D_0, and let $m > 2D_0$. Let τ be the time delay. Assume that A contains only a finite number of equilibria (i.e., fixed states of the dynamical system) and a finite number of periodic orbits of period $p\tau$ for $3 \leq p \leq m$ and that there are no periodic orbits of period τ of 2τ. Then, with probability 1, a choice of measurement function h yields a delay coordinate function $\mathbf{H}$ that is one to one from A to $\mathbf{H}(A)$.

The one-to-one property is guaranteed to fail not only when the sampling rate is equal to the frequency of a periodic orbit, as discussed above, but also when the sampling rate is twice the frequency of a periodic orbit, that is, when A contains a periodic orbit of minimum period 2τ. To see that this is so, define the function $\eta(\mathbf{x}) = h(\mathbf{x}) - h(\phi_{-\tau}(\mathbf{x}))$ on the periodic orbit, where ϕ_t denotes the action of the dynamics over time t. The function η is either identically zero or it is nonzero for some $\mathbf{x}$ on the periodic orbit, in which case it has the opposite sign at the image point $\phi_{-\tau}(\mathbf{x})$ and changes sign on the periodic orbit. In any case, $\eta(\mathbf{x})$ has a root $\mathbf{x}_0$. Since the period is 2τ, we have $h(\mathbf{x}_0) = h(\phi_{-\tau}(\mathbf{x}_0)) = h(\phi_{-2\tau}(\mathbf{x}_0)) = \cdots$. Then the delay coordinate map $\mathbf{x}_0$ and $\phi_{-\tau}(\mathbf{x}_0)$ are distinct, so $\mathbf{H}$ is not one to one for any observation function h. This problem may be eliminated by proper choice of τ.

Unfortunate choices of τ that are not ruled out by the theory are those that are unnaturally small or large in comparison to the time constant of the system. Such values of τ will cause the correlation between successive measurements to be excessively large or small, causing the effectiveness of the reconstruction to degrade in real-world cases, where noise is present.

The choice of optimal time delay for unfolding the reconstructed attractor is an important and largely unresolved problem. A commonly used rule of thumb is to set the delay to be the time lag required for the autocorrelation function to become negative (zero crossing) or, alternatively, the time lag required for the autocorrelation function to decrease by a factor of e. Another approach, that of Fraser and Swinney [1986], incorporates the concept of mutual information, borrowed from Shannon's information theory, which provides a measure of the general independence of two variables. They suggest choosing the time delay that produces the first local minimum of the mutual information of the observed quantity and its delayed value. Other aspects of the problem have been clarified by Liebert and Schuster [1988].

Finally, although we have discussed using simultaneous measurements and delay coordinates separately, there is no theoretical restriction against mixing the two ideas. Analogous theorems can be proved that allow one-to-one reconstruction of the attractor with m_1 delay coordinates and m_2 independent simultaneous coordinates as long as $m_1 + m_2 > 2D_0$.

8.2.2 Differentiable Embedding

Assume that A is a compact smooth d-dimensional submanifold of R^k. A circle is an example of a smooth one-dimensional manifold; a sphere and torus are examples of two-dimensional manifolds. A smooth d-manifold has a well-defined d-dimensional tangent space at each point. If **F** is a smooth function from one manifold to another, then the Jacobian matrix **DF** maps tangent vectors to tangent vectors. More precisely, for each point **x** on A, the map **DF(x)** is a linear map from the tangent space at **x** to the tangent space at **F(x)**. If for all **x** in A no nonzero tangent vectors map to zero under **DF(x)**, then **F** is called an *immersion*.

Figure 8.5*a* shows an immersion of the circle A. In Fig. 8.5*b*, although **F** is one to one on A, the function **F** fails to be an immersion at the pinch point. Figure 8.5*c* shows an immersion that is not a topological embedding.

A smooth function **F** on a smooth manifold is called a *differentiable embedding* if **F** and $\mathbf{F}^{-1}$ are one-to-one immersions. In particular, a differentiable embedding is automatically a topological embedding. In addition, the tangent spaces of A and **F**(A) are isomorphic. In particular, the image **F**(A) is also a smooth manifold of the same dimension as A.

In 1936, Whitney proved that if A is a smooth d-manifold in R^k, and if $m > 2d$, then a typical map from R^k to R^m is a differentiable embedding when restricted to A. Takens [1981] proved a result in this context for delay coordinate functions. The following is a version of Takens's theorem by Sauer et al. [1993].

> *Differentiable Embedding—Delay Coordinates:* Assume that a continuous-time dynamical system has a compact invariant smooth manifold A of dimension d, and let $m > 2d$. Let τ be the time delay. Assume that A contains only a finite number of equilibria, no periodic orbits of periods τ or 2τ, and only a finite number of periodic orbits of period $p\tau$ for $3 \le p \le m$. Assume that the Jacobian of the return maps of those periodic orbits have distinct eigenvalues. Then, with probability 1, a choice of the measurement function h yields a delay coordinate function **H** that is a differentiable embedding from A to **H**(A).

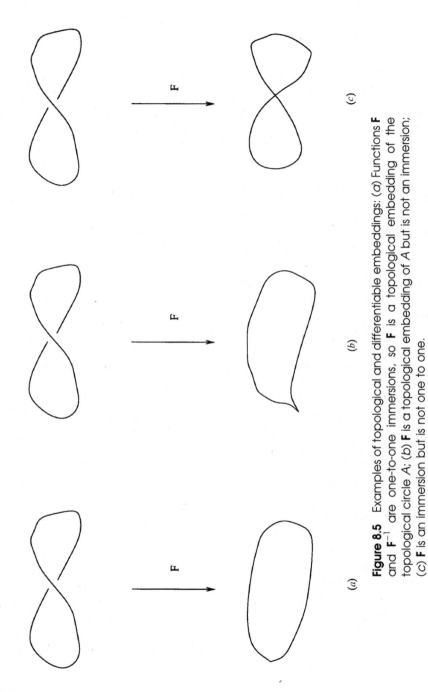

Figure 8.5 Examples of topological and differentiable embeddings: (*a*) Functions **F** and **F**$^{-1}$ are one-to-one immersions, so **F** is a topological embedding of the topological circle *A*; (*b*) **F** is a topological embedding of *A* but is not an immersion; (*c*) **F** is an immersion but is not one to one.

This version differs from Takens's original statement in two ways: Instead of assuming the generic nature of the dynamical system, the requirements on the system (in particular, the relation of periodic orbits to the time delay) are made explicit, and "generic" is replaced by "probability 1."

From the point of view of extracting information from an observed dynamical system, there are some advantages to having a differentiable, as compared with topological, embedding. These advantages stem from the fact that metric properties are preserved by the reconstruction.

First, because a differentiable embedding is a C^1-diffeomorphism from the phase-space attractor to the reconstructed attractor, there is a uniform upper bound on the stretching done by both $\mathbf{H}$ and $\mathbf{H}^{-1}$. This is in contrast to the topological embedding case. One typically knows that $\mathbf{H}$ is continuously differentiable, giving an upper bound on the stretching of $\mathbf{H}$ on a compact attractor. However, in general, $\mathbf{H}^{-1}$ can be nondifferentiable, even though $\mathbf{H}$ is one to one. Functions with an upper bound on stretching are called *Lipschitz*; if $\mathbf{H}$ and $\mathbf{H}^{-1}$ are Lipschitz, then $\mathbf{H}$ is bi-Lipschitz. Under virtually any reasonable definition of dimension that measures metric information, the dimension of a set is unchanged under a bi-Lipschitz map. Therefore, the hypotheses of statement above imply that the fractal dimensions of every closed subset of A can be calculated in principle by delay coordinate embedding.

Of course, in order to investigate the dimension of an arbitrary set S, we first need to find a d-dimensional differentiable manifold A that contains S and satisfies those hypotheses and to use at least $2d$ delay coordinates. In many cases, no natural manifold structure exists for an attractor A. In those cases, it is necessary to rely on a separate theorem (see Ding et al., 1993), which says that, in general, the correlation dimension of the reconstructed attractor $\mathbf{H}(A)$ in R^m equals the correlation dimension of A as long as the number of delay coordinates m is greater than the correlation dimension of A. This fact is independent of whether $\mathbf{H}$ is one to one or not.

A second advantage of differentiable embedding is that all Lyapunov exponents of the phase-space attractor are reproduced in the reconstruction. If ϕ_t denotes the dynamical flow on the phase-space manifold A, then $\psi_t \equiv \mathbf{H}\phi_t\mathbf{H}^{-1}$ defines the flow in the reconstructed manifold. Since $\mathbf{H}$ and $\mathbf{H}^{-1}$ are differentiable, the chain rule $\mathbf{D}\psi_t = \mathbf{DHD}\phi_t\mathbf{DH}^{-1}$ shows that $\mathbf{D}\psi_t$ and $\mathbf{D}\phi_t$ are similar matrices. As a result, the d-Lyapunov exponents of the dynamics on A will be reproduced on the manifold $\mathbf{H}(A)$. In practice, the reconstructed manifold will be embedded as a submanifold of R^m, so there will be $m - d$ "spurious" Lyapunov exponents not associated with the dynamics on A that are artifacts of the embedding process. Distinguishing the spurious from the true Lyapunov exponents is an important task and is done by local false-nearest-neighbor analysis, which is discussed later.

8.3   PHASE-SPACE RECONSTRUCTION

From a signal-processing perspective, an issue of paramount importance is the reconstruction of dynamics from measurements made on a single coordinate of the system. The motivation here is to make "physical sense" from the resulting time series, bypassing a detailed mathematical knowledge of the underlying dynamics.

Figure 8.6 gives a schematic description of a phase-space reconstruction for the Lorenz map in the chaotic regime. Figure 8.6a is the projection of the trajectory on the (x, y) plane. Figure 8.6b is the scalar time series corresponding to the time evolution of the x-variable, and Fig. 8.6c is the reconstructed trajectory from the x-variable in the $(x(n), x(n + \tau))$ plane. There is a differentiable equivalence between trajectories in Figs. 8.6a and 8.6b.

Let the time series be denoted by $s(nT)$, $n = 0, 1, \ldots, N$, where N is the total number of samples and T is the sampling period. To reconstruct the dynamics of the original attractor that gave rise to the observed time series, we seek an embedding space where we may reconstruct an attractor from the scalar data so as to preserve the invariant characteristics of the original unknown attractor [Schuster, 1988]. To simplify the mathematical details, we set $T = 1$. By applying the delay-coordinate method [Haykin and Leung, 1992], the $d_E \times 1$ phase-space vector $\mathbf{s(n)}$ is constructed by assigning coordinates:

$$
\begin{aligned}
s_1(n) &= s(n) \\
s_2(n) &= s(n - \tau) \\
&\vdots \\
s_{d_E}(n) &= s(n - (d_E - 1)\tau)
\end{aligned}
\tag{8.6}
$$

where d_E is the *embedding dimension* and τ is the *normalized embedding delay*. These parameters are not chosen arbitrarily; rather they have to be

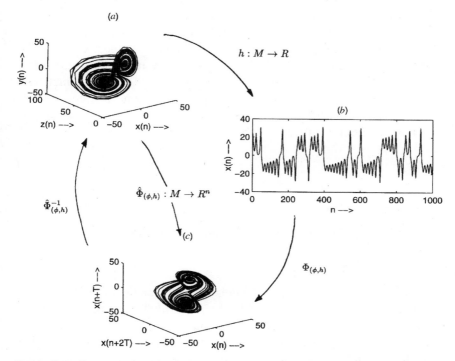

Figure 8.6   Schematic description of phase-space reconstruction for Lorenz attractor. (*a*) Lorenz attractor, (*b*) *x*-component of the Lorenz map, (*c*) Reconstructed Lorenz attractor from *n*-component alone using delay-coordinate embedding.

determined experimentally. Procedures for estimating the normalized embedding delay (τ) based on *mutual information* (MI) are described in the literature [Fraser, 1986; Fraser and Swinney, 1986; Pineda and Sommerer, 1994]. For estimating the embedding dimension d_E, we may use the method of *global false nearest neighbor* (GFNN) [Abarbanel, 1996; Abarbanel et al., 1993; Kennel et al., 1992; Liebert et al., 1991].

Another important parameter that needs to be determined in the analysis of an observed chaotic process is the *local* or *dynamical dimension* (d_L) [Abarbanel and Kennel, 1993]. The local dimension represents the number of dynamical degrees of freedom that are active in determining the evolution of the system as it moves around the attractor

[Abarbanel, 1996; Abarbanel et al., 1993]. This integer dimension gives the number of true Lyapunov exponents of the system under investigation. It is less than or equal to the global embedding dimension; that is, $d_L \leq d_E$. The parameter d_L can be determined by using the *local false-nearest-neighbor* (LFNN) method [Abarbanel, 1996; Abarbanel and Kennel, 1993; Abarbanel et al., 1993]. The notion of false nearest neighbors refers to nearest neighbors as an attractor that have become near one another through projection when the attractor is viewed in a dimension too low to unfold the attractor completely.

8.4 ESTIMATION OF EMBEDDING PARAMETERS

8.4.1 Embedding Delay

In principle, any choice of the time delay τ is acceptable in the limit of an infinite amount of noiseless data. In the more likely situation of a finite amount of data, the choice of τ is of *considerable* practical importance in trying to reconstruct the attractor that represents the dynamical system that generated the data. The choice of time delay should satisfy the following for a time-delay embedding (Abarbanel, 1996):

- It must be some multiple of the sampling time T, since we only have data at those times.
- If the time delay is too short, the coordinates $s(n)$ and $s(n + \tau)$ that we wish to use in our reconstructed data vector $\mathbf{s(n)}$ will not be independent enough. That is to say, not enough time will have evolved for the system to have explored enough of its state space to produce, in a practical numerical sense, new information about that state space.
- Since chaotic systems are intrinsically unstable, if τ is too large, any connection between the measurements $s(n)$ and $s(n + \tau)$ is numerically tantamount to being random with respect to each other.

The time delay may be estimated using the autocorrelation method or the mutual information method. These two methods are described briefly in the following subsections.

Autocorrelation Function

The autocorrelation function of the sampled data set $x_i = x(t_0 + iT)$, where T is the sampling period and $i = 1, 2, \ldots, N_0$, is given by

$$C(\tau) = \frac{\sum_{k=1}^{N_0} [x(t_0 + kT + \tau) - \bar{x}][x(t_0 + kT) - \bar{x}]}{\sum_{k=1}^{N_0} [x(t_0 + kT) - \bar{x}]^2} \tag{8.7}$$

where $\bar{x}$ is the mean defined by

$$\bar{x} = \frac{1}{N_0} \sum_{k=1}^{N_0} x(t_0 + kT) \tag{8.8}$$

Then, if the autocorrelation function has a zero crossing at τ, the corresponding value of time delay is chosen to be the optimum time delay for time-delay embedding. Otherwise, the first local minimum of the autocorrelation function is used to specify the optimum delay.

Note that the autocorrelation function provides only a linear measure of the independence between the coordinates $x(t_0 + k\tau_s + \tau)$ and $x(t_0 + k\tau_s)$. Researchers have shown that the time delay estimated using this method may not be the optimum one for nonlinear time series.

Mutual Information Method

The optimum time delay (τ) can be estimated by looking at the mutual information, which is an information-theoretic property derived from entropy considerations [Shannon and Weaver, 1949]. It determines how much information the measured signal $s(n)$ has relative to the measurements at some other time $s(n + \tau)$. The mutual information between $s(n)$ and $s(n + \tau)$ is formally defined as

$$I(\tau) = \sum_{s(n),s(n+\tau)} P(s(n),\ s(n + \tau)) \log_2 \left[\frac{P(s(n),\ s(n + \tau))}{P(s(n))P(s(n + \tau))} \right] \tag{8.9}$$

where $P(s(n))$ and $P(s(n + \tau))$ are the normalized distributions of $s(n)$ and $s(n + \tau)$ as observed and $P(s(n),\ s(n + \tau))$ is the joint distribution of the two measurements.

The parameter $I(\tau)$ is always greater than or equal to zero [Gallager, 1968]. So, we cannot look for a zero of the mutual information in analogy with familiar linear correlation. Generally, for attractor reconstruction, a good practice for selecting τ is to use the first minimum of $I(\tau)$ [Fraser and Swinney, 1986].

In general, the prescription provided by the mutual information often appears to yield a shorter time delay compared to the first zero crossing of the autocorrelation function [Abarbanel, 1996; Abarbanel et al., 1998]. The problem in using the longer time delay is that the components of the vector may become independent of each other, causing subsequent calculations to be invalid [Abarbanel et al., 1998].

8.4.2 Global Embedding Dimension (d_E)

Determining an optimum embedding dimension is another important step in phase-space reconstruction. The purpose of time-delay embedding is to unfold the projection back to a multivariate state space that is a topological representation of the original system [Landa and Rosenblum, 1991]. The theorems of Takens [1981] and Mañé [1981] state that an attractor of dimension d_A can be unfolded in a space whose dimension d_E satisfies the relation $d_E \geq 2d_A + 1$. Later, Sauer found that it is sufficient to satisfy the relation $d_E \geq d_A$ for proper embedding [Sauer et al., 1991]. In 1992, Kennel et al. [1992] showed that the Mañé [1981] and Takens [1981] theorems are only sufficient conditions, and they showed that the familiar Lorenz attractor ($D_2 = 2.06$) can be embedded in a three-dimensional ($d_E = 3$) embedding space using the time-delay method [Brown et al., 1991; Bryant et al., 1990]. This is in contrast to Takens's theorem where a dimension of $d_E = 5$ has to be used for embedding a Lorenz system. Kennel et al. [1992] pointed out that the use of excess dimensionality unnecessarily increases the computational complexity and enhances chances of contamination by roundoff or instrument errors. They proposed a false-nearest-neighbor method for the selection of optimum embedding dimension [Brown et al., 1991; Bryant et al., 1990].

The method of false nearest neighbors relies on the geometric basis for the embedding theorem: As the dimension is increased, attractors unfold. Points on trajectories that appear close in dimension $d(< d_E)$ may move apart in dimension $d + 1$. These are "false" neighbors in d dimension, and the method measures the percentage of false neighbors as d increases. Let $\mathbf{s}^{NN}(k) = [s^{NN}(k), s^{NN}(k-1), \ldots, s^{NN}(k + (d-1)\tau]^{\mathrm{T}}$ be the nearest neighbor of $\mathbf{s}(k) = [s(k), s(k-1), \ldots, s(k + (d-1)\tau)]^{\mathrm{T}}$ in $d - 1$ dimension. This neighbor is false in dimension d if

$$\frac{R_d^2(k)}{R_{d-1}^2(k)} > R_{\text{toll}} \tag{8.10}$$

where $R_d^2(k)$ is the Euclidean distance between a point $s(k)$ and $s^{NN}(k)$ and R_{tol1} is the criterion for declaring whether the neighbors that are close in d are distant in $d + 1$.

A second criterion is necessary because the nearest neighbor may not necessarily be "close." The density of vectors in space may be low as the dimension increases. That is, as dimensions are added, the proportionate volume occupied by the signal will decrease and the distance to neighbors will increase. If the nearest neighbor to a point is false but not close, then the Euclidean distance in going to $d + 1$ will be $\approx 2R_A$. So, the second criterion is

$$\frac{R_d^2(k)}{R_A^2} > R_{tol2} \tag{8.11}$$

where

$$R_A^2 = \frac{1}{N} \sum_{k=1}^{N} [s(k) - \bar{s}]^2 \tag{8.12}$$

and

$$\bar{s} = \frac{1}{N} \sum_{k=1}^{N} [s(k)] \tag{8.13}$$

A nearest neighbor is false if either of these two tests fails. For a noise-free signal, the number of false nearest neighbors becomes zero when the minimum embedding dimension d_E is reached. In the case of a noisy signal, it may drop off dramatically but may not become zero [Abarbanel, 1996; Abarbanel et al., 1993; Brown et al., 1991; Bryant et al., 1990].

8.4.3   Local Embedding Dimension (d_L)

Once we have determined the integer global dimension (d_E) required to unfold the attractor on which the data reside, we still have the question of the number of dynamical degrees of freedom that are active in determining the evolution of the system as it moves around the attractor. The number of active degrees of freedom is somewhat an intuitive concept and is referred to as the *local dimension* (d_L). Clearly, $d_L \leq d_E$; d_L gives the number of true Lyapunov exponents in the system under investigation. All the important dynamical behaviors can be captured in d_L. If $d_L < d_E$, then the model of the dynamics can be simpler than one would conclude from the global dimension required to unfold the entire attractor.

The method for calculating the local dimension is straightforward. Trajectories associated with true neighbors ought to remain close for some period of time before they diverge due to the instability associated with the positive Lyapunov exponents of the dynamics. If the neighboring trajectories separate faster than is expected, it can be presumed that the separation is caused by the projection onto too few dimensions, where neighbors may be false dynamically due to the projection alone. The task is simply to define a criterion for how fast sufficiently embedded trajectories should separate and then test a large number of neighborhoods against this criterion.

The first thing to be addressed in this analysis is to decide upon the coordinate system in which to ask the question of local false neighbors. The best approach is to work in a global dimension that is large enough to assure that all neighbors are true, namely, some working dimension d_W such that $d_W \geq d_E$. In this space, we choose a point $\mathbf{s}(k)$ on the attractor and find what subspace of dimension $d_L \leq d_E$ allows one to make accurate local neighborhood-to-neighborhood maps of the data on the attractor. To do this, we must define a neighborhood, which may be done by specifying the number of neighbors N_B of the point $\mathbf{s}(k)$ and then providing a local rule for how these points evolve in one time step into the same N_B points near $\mathbf{s}(k + 1)$. Then, we choose that value of d_L where the quality of prediction becomes independent of both d_L and the number of nearest neighbors N_B.

The choice of d_L is based on principal-component analysis of the data [Gollub and VanLoan, 1989]. This method locally selects those directions in d_W-dimensional space that contain the majority of the data in the least-squares sense. This local principal-component decomposition in dimension d_W may be performed by forming the sample covariance matrix among the N_B neighbors $\mathbf{s}^{(r)}(k)$; $r = 1, 2, \ldots, N_B$ of $\mathbf{s}(k)$. The sample covariance matrix is the $d_W \times d_W$ matrix

$$\mathbf{R}(k) = \frac{1}{N_B} \sum_{r=1}^{N_B} [\mathbf{s}^{(r)}(k) - \bar{\mathbf{s}}(k)][\mathbf{s}^{(r)}(k) - \bar{\mathbf{s}}(k)]^{\mathrm{T}} \tag{8.14}$$

where

$$\bar{\mathbf{s}} = \frac{1}{N_B} \sum_{r=1}^{N_B} [\mathbf{s}^{(r)}(k)] \tag{8.15}$$

The eigenvalues of this covariance matrix are ordered by size; the local d_L dimensional space is chosen as the eigenvectors associated with the d_L largest eigenvalues.

Once the d_L d_W-dimensional basis vectors are obtained, we perform the projection of these d_W-dimensional vectors $\mathbf{s}^{(r)}(k)$ onto the d_L eigen-directions. These vectors are called $\mathbf{z}^{(r)}(k)$ and constitute the N_B local d_L-dimensional vectors at "time" k. Next, we find the vectors $\mathbf{z}^{(r)}(k; \Delta)$ that evolve from the $\mathbf{z}^{(r)}(k)$ in Δ time steps. To do this, we construct a local polynomial map

$$\mathbf{z}^{(r)}(k) \rightarrow \mathbf{z}^{(r)}(k; \Delta)$$
$$\mathbf{z}^{(r)}(k; \Delta) = \mathbf{A} + \mathbf{B}\mathbf{z}^{(r)}(k) + \mathbf{C}\mathbf{z}^{(r)}(k) + \cdots \tag{8.16}$$

which takes the vectors $\mathbf{z}^{(r)}(k)$ into their counterparts $\mathbf{z}^{(r)}(k; \Delta)$. The coefficients $\mathbf{A}$, $\mathbf{B}$, and $\mathbf{C}$ are determined in the least-squares sense. We should look only at the local linear maps, since the operations are taking place locally.

Having determined the local linear map, we check how well it predicts forward in time. In particular, we should check when the local polynomial prediction map fails, because it makes a prediction error in Δ steps that is some finite fraction ($0 \leq \beta \leq 1.0$) of the size of the attractor:

$$\mathbf{R}_A = \frac{1}{N} \sum_{k=1}^{N} |s(k) - \bar{s}(k)| \tag{8.17}$$

as defined above. When the percentage of bad predictions becomes independent of d_L and is also insensitive to the number of nearest neighbors N_B, we may conclude that the correct local dimension for the active dynamical degrees of freedom has been reached.

8.5 RESULTS OF ESTIMATION OF EMBEDDING PARAMETERS

The determination of embedding parameters (i.e., the embedding dimensions d_E and normalized embedding delay τ) is an essential step before performing the statistical tests on a chaotic time series. These parameters are obtained by using the GFNN and MI algorithms, respectively [Abarbanel, 1996; Abarbanel et al., 1993; Brown et al., 1991; Fraser, 1986; Fraser and Swinney, 1986; Kennel et al., 1992; Liebert et al., 1991]. Figure 8.7 shows the results obtained using the MI algorithm for one particular clutter data set. Here we used an embedding dimension $d_E = 2$ and the normalized embedding delay varied from 0 to 50. The optimum embedding delay τ is 11, where the MI reaches a minimum for the first time. Note that the corresponding minimum value of the MI is small, which indicates that radar samples spaced by $\tau = 11$ time units apart are essentially independent; and yet it is nonzero, which indicates that the radar samples so delayed are correlated with each other. The optimum embedding dimension for the same data was found to be $d_E = 5$ from the GFNN analysis. Figure 8.8 shows the percentage of GFNN versus different embedding dimensions. We used thresholds R_{tol1} and R_{tol2} of 25 and 10, respectively, and a normalized embedding delay $\tau = 11$ (obtained from the MI analysis) for this calculation. This procedure was repeated for the entire database used for the study.

The LFNN analysis for estimating the local embedding dimension (d_L) of sea clutter data was performed for about 30 data sets drawn from different categories of the data sets used for the study. The analysis included data sets from different wave heights of sea surface, raw (I–Q corrected), three-point smoothed and FIR filtered in-phase, quadrature-phase, and amplitude components from different radar sites. Results of this study show that the local dimension varies between 5 and 6. However, in the majority of cases the local dimension was 5. Figure 8.9 illustrates the results of the local false-nearest-neighbor analysis on the same clutter data used in Fig. 8.8. The four curves shown in Fig. 8.9 correspond to local nearest neighbors (N_B) of 40, 60, 80, and 100. It may be observed that beyond an embedding dimension $d_L = 5$ the percentage of bad predictions is almost independent of both the choice of d_L and the choice of N_B. Figure 8.9 implies that a local dimension $d_L = 5$ is adequate for the analysis; this result defines the number of true Lyapunov exponents. Note also that for the same data set used in Fig. 8.8 we have $d_E = 5$, which is in agreement with the requirement that $d_E \geq d_L$.

8.5.1 Comparison of d_E for Sea Clutter, Tides, and Correlated Noise

In this section we present a comparison of global embedding dimensions for three different processes: sea clutter, tides, and colored noise. This exercise is to authentically show the distinction between sea clutter dynamics and colored noise.

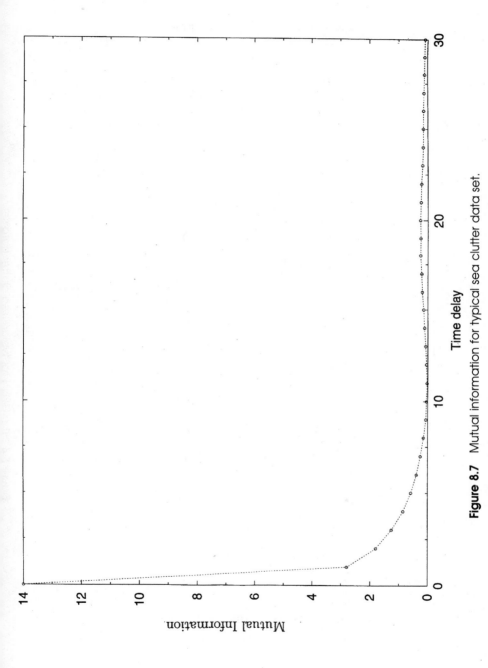

Figure 8.7 Mutual information for typical sea clutter data set.

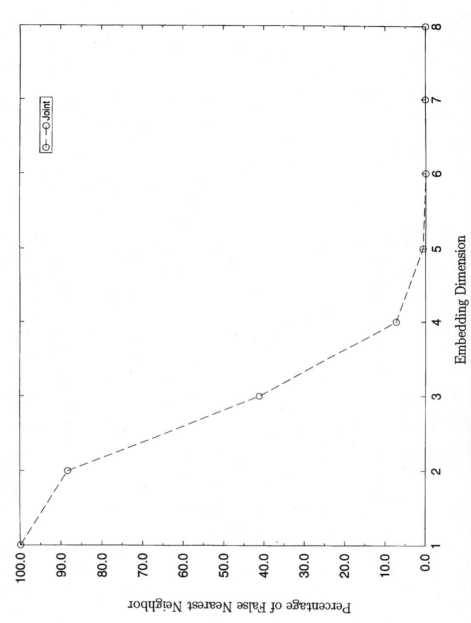

Figure 8.8 The GFNN for typical sea clutter data set

116

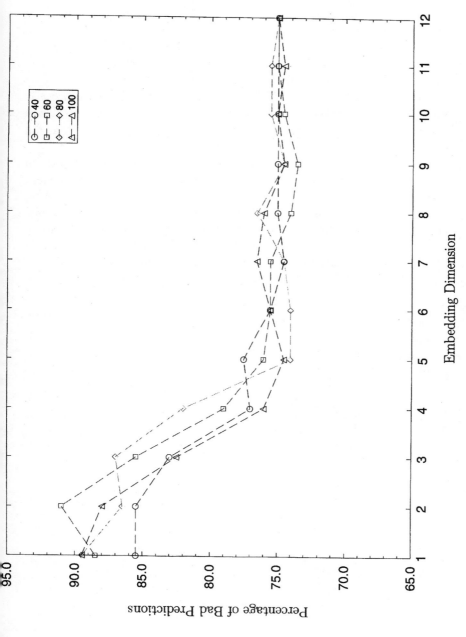

Figure 8.9 The LFNN for same sea clutter data used in Fig. 8.8.

117

Correlated noise, in general, has a broadband spectrum that is similar to that of the power spectrum of a chaotic process. What it implies is that a distinction between these two types of processors cannot be made by simply looking at their power spectra. In other words, processors that have similar power spectra need not have originated from the same source. This is what we showed with our WSF experiment described in Section 6.3.1. In this section we use the global embedding dimensions of these processes to distinguish the deterministic origin of sea clutter and tides from the random origin of the colored-noise counterpart.

In Section 6.3, we performed the method of surrogate data analysis to discriminate between a signal generated by a deterministic chaotic system (sea clutter) and colored noise. We performed the estimation of chaotic invariants on both the filtered noise and the actual sea clutter for comparison and showed that their invariants are really distinct. Noisy process, by definition, is an infinite-dimensional process. By the method of GFNN we are trying to find an optimum embedding dimension for the underlying process. We showed that this dimension converges to 5 for sea

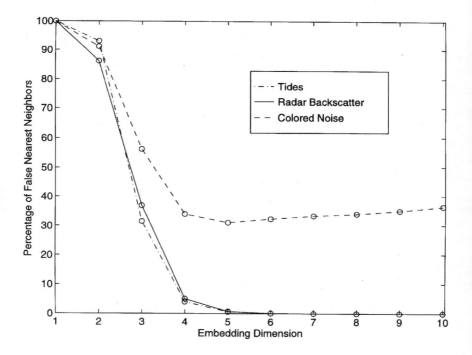

Figure 8.10   Results for GFNN analysis performed on different data sets.

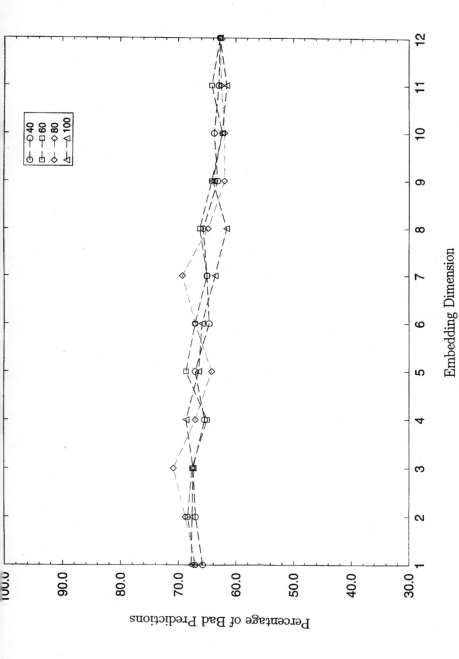

Figure 8.11 The LFNN for colored noise.

clutter data in the majority of our studies. Results of the GFNN analysis performed on the filtered noise are plotted alongside the corresponding results for the sea clutter data, as shown in Fig. 8.10. We can clearly see from Fig. 8.10 that the percentage of FNN converges to zero beyond embedding dimension 5 for sea clutter data, whereas it behaves quite differently for the noisy process. Figure 8.10 also includes the GFNN results for tides data. This later experiment was performed by Abarbanel et al., 1998]. (We plotted the curve of percentage of FNN by taking coordinate points from Fig. 8 of the article from Abarbanel et al., 1998.)

The local dimensionality, which gives the active degrees of freedom for the underlying process, has been computed for both the sea clutter data and the colored noise. The LFNN analysis performed on the sea clutter data is shown in Fig. 8.8, and the corresponding results for the colored noise are presented in Fig. 8.11. From these two pictures it is clear that for sea clutter data there is clearly an active degree of freedom $d_L = 5$, whereas for the colored noise it is undefined as the percentage of bad predictions never converged.

The experimental results presented in this section show that sea clutter is generated by a deterministic nonlinear process and should not be misinterpreted as a colored-noise process.

9

ESTIMATION OF CHAOTIC
INVARIANTS

9.1 INTRODUCTION

Characterization of dynamical systems is an important part of the analysis of observed signals. As mentioned in Chapter 4, the important characteristic invariants of a chaotic process are (1) correlation dimension or fractal dimension (D_2), (2) Lyapunov exponents (λ_i) and hence the Lyapunov dimension or Kaplan–Yorke dimension (D_{KY}), and (3) Kolmogorov entropy (K).

In this chapter we describe robust procedures for the estimation of these chaotic invariants and related parameters. Most importantly, they will be applied to real-life sea clutter data collected under varying environmental conditions. Bearing in mind the title of the book, the results presented in this chapter are central to the whole book.

9.2 ESTIMATION OF CORRELATION DIMENSION

The correlation dimension (D_2) has been the most intensely studied invariant quantity for dynamical systems [Abarbanel, 1996; Abarbanel et al., 1993; Baker and Gollub, 1996; Grassberger and Procaccia, 1983; Mandelbrot, 1983; Moon, 1992; Osborne and Provenzale, 1989; Ott,

1993; Provenzale et al., 1992; Schuster, 1988; Theiler, 1988, 1990a,b]. Unlike the integer Euclidean dimension, D_2 has a fractal value [Abarbanel, 1996; Abarbanel et al., 1993; Baker and Gollub, 1996; Mandelbrot, 1983; Moon, 1992; Ott, 1993; Schuster, 1988]. This dimension of a dynamical system directly reflects the way in which points on the attractor are distributed in the phase space or embedding space [Grassberger and Procaccia, 1983]. The parameter D_2 reflects the complexity of the underlying dynamical system and bounds the degrees of freedom or the number of parameters required to describe the system under investigation. It is estimated from the experimental time series by looking at the way in which points within a sphere of radius r scales as the radius shrinks to zero in the embedding space [Grassberger and Procaccia, 1983]. There are already a large number of very good reviews of methods for the estimation of D_2 in the literature [Abarbanel, 1996; Abarbanel and Sushchik, 1993b; Abarbanel et al., 1993; Broomhead and King, 1986; Diks, 1996; Ding et al., 1993; Eckmann and Ruelle, 1992; Ellner, 1988; Farmer et al., 1983; Fukunga and Olsen, 1971; Grassberger, 1990; Grassberger and Procaccia, 1983; Hedigar et al., 1990; Henderson and Wells, 1988; Osborne and Provenzale, 1989; Paladin and Vulpiani, 1987; Provenzale et al., 1992; Russell et al., 1980; Schouten et al., 1994; Takens, 1983b; Theiler, 1988, 1990a,b].

A classical approach to the estimation D_2 was provided by Grassberger and Procaccia [1983]. Their algorithm, popularly known as the GPA, computes the number of couples $(\mathbf{s}(i), \mathbf{s}(j))$ whose magnitude is less than a given radius r in the multidimensional space [Grassberger and Procaccia, 1983a,b]. This is pictured in Fig. 9.1. More precisely, the GPA computes the correlation integral $C(r)$ given by

$$C(r) = \frac{1}{N(N-1)} \sum_{i=1}^{N} \sum_{\substack{j=1 \\ j \neq i}}^{N} \theta(r - \|\mathbf{s}(i) - \mathbf{s}(j)\|) \qquad (9.1)$$

where θ is the Heaviside function, $\mathbf{s}(i)$ and $\mathbf{s}(j)$ are vectors in the multidimensional space, and N is the total number of vectors in the multidimensional space. The correlation integral obeys the following scaling law:

$$\lim_{\substack{r \text{ small} \\ N \to \infty}} C(r) \approx r^{D_2} \qquad (9.2)$$

from which we can calculate D_2 as the slope of log–log curves as shown by

$$D_2 = \lim_{r \text{ small}} \frac{\log[C(r)]}{\log(r)} \qquad (9.3)$$

Here, $C(r)$ is a measure of the probability that two arbitrary points $s(i)$ and $s(j)$ of the multidimensional space will be separated by a distance r. The main point is that $C(r)$ behaves as a power of r for small values of r. Therefore, plotting $\log[C(r)]$ versus $\log(r)$ allows us to calculate D_2 from the slopes of the curves. If the slopes of the curves for increasing embedding dimension converge to a saturation value, this limit is called the correlation dimension (D_2). For a stochastic process, there is no convergence and the slopes keep on increasing with an increase in the embedding dimension.

The algorithm developed by Grassberger and Procaccia [1983] is perhaps the most widely used one for computing D_2. Basically, the GPA measures the rate of change of the local point densities around the attractor

Multi-dimensional Space

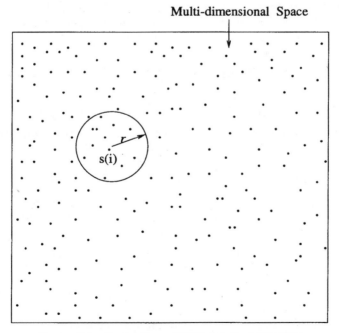

Figure 9.1   Illustration of GPA for D_2 estimation.

surface as a function of the local radius. However, this algorithm suffers from certain disadvantages: (1) highly demanding computational requirements [Eckmann and Ruelle, 1992; Essex and Narenberg, 1991; Ruelle, 1990] and (2) spurious estimates of D_2 due to the presence of measurement noise. For a data length N, the GPA needs a computation $O(N^2)$ for one local center and one embedding dimension. This has to be repeated for more and more embedding dimensions. Moreover, a reliable estimate of D_2 using the GPA requires a large amount of noise-free data. The GPA fares poorly in the presence of noise (for SNR less than 20 dB), where the plot of $\log[C(r)] - \log(r)$ may not even converge. Further, the scaling region for determining the slope (and hence D_2) is more often arbitrarily fixed, thereby giving rise to varying results by different investigators [Ding et al., 1993].

These limitations of the GPA motivated us to use a more reliable method, namely, one based on the maximum-likelihood principle for the estimation of D_2 [Ellner, 1988; Takens, 1983b,c]; maximum likelihood is regarded in the statistics literature as a reliable method for parameter estimating. Schouten et al., [1994] have developed such a method for estimating D_2 in the presence of noise. We refer to this algorithm as the STB$_1$ algorithm in recognition of its originators, Schouten, Takens, and Bleek.

9.2.1 Maximum-Likelihood Estimate of the Correlation Dimension

This method is based on the correlation integral that is defined in Eq. (9.1). Choose an upper bound r_0 on the scaling distance r because of the fact that the scaling relationship holds only for sufficiently small r. For large r, the correlation integral $C(r)$ is dominated by saturation effects $[C(r) \to 1]$ when r reaches the order of the size (horizontal extent) of the attractor. Also, $r_0 > r_n$ (where r_n is the maximum possible noise distance between two points on the noisy attractor). The distance between two points $\mathbf{s}(i)$ and $\mathbf{s}(j)$ on the attractor is defined as

$$\|\mathbf{s}(i) - \mathbf{s}(j)\| = \max_{0 \le k \le d_E - 1} |s(i+k) - s(j+k)| \qquad (9.4)$$

Influence of Noise

Let $\mathbf{s}(i)$ and $\mathbf{s}(j)$ be points located on the reconstructed attractor on different orbits. If these points are not disturbed by noise, they form the true points satisfying the dynamics of the system. The maximum norm distance between these points is then calculated by the above equation. Now let each point in the time series be corrupted by noise that is bounded in magnitude with maximum possible amplitude of $\pm\frac{1}{2}\delta s_{max}$. In that case, the elements $z(i, k)$ and $z(j, k)$ of the noisy vector $\mathbf{z}(i)$ and $\mathbf{z}(j)$ are each assumed to be composed of a noise-free part $s(i, k)$ and $s(j, k)$ and a noise factor $\delta s(i, k)$ and $\delta s(j, k)$ according to

$$z(i, k) = s(i, k) + \delta s(i, k) \tag{9.5}$$

and

$$z(j, k) = s(j, k) + \delta s(j, k) \tag{9.6}$$

with

$$-\tfrac{1}{2}\delta s_{max} \leq \delta s(i) \leq +\tfrac{1}{2}\delta s_{max} \tag{9.7}$$
$$-\tfrac{1}{2}\delta s_{max} \leq \delta s(j) \leq +\tfrac{1}{2}\delta s_{max} \tag{9.8}$$

We assume that there exists a trajectory satisfying the nonlinear dynamics of the chaotic system, which is sufficiently close to the measured, noise-corrupted trajectory. Theoretically, when $d_E \to \infty$, the probability of finding two corresponding elements $z(i, k)$ and $z(j, k)$ that are maximally corrupted with $-\frac{1}{2}\delta s_{max} + \frac{1}{2}\delta s_{max}$, respectively, will be *unity*. Futhermore, we need the maximally corrupted pair also to be the pair for which $|s(i, k) - s(j, k)|$ is maximal. The maximum norm distance between the corrupted vectors is

$$\begin{aligned} r_z &= \lim_{d_E \to \infty} \max_{0 \leq k \leq d_E-1} |z(i, k) - z(j, k)| \\ &= \lim_{d_E \to \infty} \max_{0 \leq k \leq d_E-1} |s(i, k) - s(j, k)| + \delta s_{max} \\ &= r_x + r_n \end{aligned} \tag{9.9}$$

where r_z is the corrupted distance, r_x is the noise-free distance, and $r_n = \delta s_{max}$ is the maximum noise distance.

From Eq. (9.9), the probability of finding interpoint distances r_z below $r_n = \delta s_{max}$ is *zero*. This means that $C(r_z \leq r_n) = 0$ and $C(r_z > r_n) > 0$. When the power law dependency holds for the noise-free distance r_x according to $C(r_x) \sim (r_x)^{D_2}$, we can write

$$C(r_z | r_z > r_n) \sim (r_z - r_n)^{D_2} \tag{9.10}$$

with the result that $C(r_z = r_n) = 0$ and $C(r_z = r_0) = 1$. We thus obtain

$$C(r_z) = \left[\frac{r_z - r_n}{r_0 - r_n}\right]^{D_2} \qquad \text{for} \quad r_n \le r_z \le r_0 \qquad (9.11)$$

Normalize all the distances with respect to r_0 using $l = r_z/r_0$ and $l_n = r_n/r_0$, in which case we may rewrite Eq. (9.11) as

$$C(l) = \left[\frac{l - l_n}{1 - l_n}\right]^{D_2} \qquad \text{for} \quad l_n \le l \le 1 \qquad (9.12)$$

Dimension Estimation

When r_n is known a priori, all distances can be rescaled. The *maximum-likelihood estimate* of the dimension of the noise-free chaotic attractor is then calculated as

$$D_{\mathrm{ML}} = \left[\frac{-1}{M} \sum_{i=1}^{M} \ln\left[\frac{l_i - l_n}{1 - l_n}\right]\right]^{-1} \qquad (9.13)$$

Here, M is the sample size of the interpoint distances l_i with $l_n \le l_i \le 1$.

The D_{ML} and D_2 are related because these invariants are obtained from the same correlation integral. The relationship is

$$D_{\mathrm{ML}} = \frac{D_2 + l_n}{1 - l_n} \qquad (9.14)$$

Thus, having calculated D_2 and then having determined l_n from a least-squares fit to the correlation integral, we can use Eq. (9.14) to calculate D_{ML}.

9.2.2 Results of Case Studies

As mentioned earlier, the correlation dimension D_2 is the most basic and static property of an attractor. We used the STB_1 algorithm for estimating the D_2 of sea clutter data [Haykin and Puthusserypady, 1997].

In this work, we made an extensive study of D_2 of sea clutter using the five data sets parameterized in Chapter 3. These data sets were collected using different radar sites, different radar parameters, and different sea states.[1] Each data set was 50,000 samples long. We throughly investigated

---

[1] *Sea state* is a single number that provides a gross summary of the degree of agitation of the sea and the characteristics of its surface; sea state and wave height are related in some fashion [Nathanson, 1969].

the effects of the following:

- Radar component: in-phase, quadrature phase, and amplitude
- Pulse duration
- Radar range
- Pulse repetition frequency
- Like polarization (HH or VV)
- Sea state

In each case, the following preprocessings were performed before extracting the chaotic parameters:

- I–Q calibration
- Three-point smoothing
- FIR filtering

The embedding parameters d_E and τ of the sea clutter data were obtained using the GFNN and MI algorithms, respectively [Schouten et al., 1994]. The D_2 values for different data sets obtained using the STB$_1$ algorithm show that the value varies between 4.1 and 4.5, irrespective of changes in the environmental conditions and radar parameters. It gives slightly higher estimates of D_2 in the case of unfiltered data where the SNR is low, which is to be expected because of the presence of noise. Apart from this fact, the estimated values were consistent for wide variations in sea clutter data. Tables 9.1–9.3 show the D_2 values for different ranges of I–Q corrected, three-point smoothed, and FIR filtered in-phase, quadrature-phase, and amplitude conponents of clutter data using the Dartmouth database. The maximum-likelihood estimate of D_2 and the Kaplan–Yorke dimension are shown side by side for each data set. The pulse duration values for these three tables are 2000, 200, and 200 ns, respectively. We observe that the pulse duration has essentially no effect on the D_2 values. Similarly, for different starting range values (1200, 1500, and 2600 m), the D_2 values do not show any dependence on range.

The effect of like polarization of the radar, HH or VV, was also evaluated. Here, too, we see that the D_2 value is essentially independent of like polarization. Table 9.1 data were obtained using horizontal like polarization, and Tables 9.2 and 9.3 data were obtained using vertical like polarization. These three tables pertain to the same radar type and the same radar site.

Table 9.1   Correlation Dimension of In-Phase, Quadrature Phase, and Amplitude of Clutter Data for Different Ranges

Range (m)	I–Q Corrected		Three-Point Smoothed		FIR Filtered	
	D_{KY}	D_{ML}	D_{KY}	D_{ML}	D_{KY}	D_{ML}
			Amplitude			
1800	4.59	4.24	4.30	4.40	4.35	4.42
2100	4.34	4.37	4.28	4.41	4.43	4.32
2400	4.60	4.42	4.30	4.52	4.48	4.45
2700	4.34	4.36	4.31	4.43	4.38	4.53
3000	4.48	4.36	4.29	4.43	4.49	4.50
			In-Phase Component			
1800	4.25	4.20	4.23	4.22	4.25	4.16
2100	4.26	4.30	4.31	4.24	4.35	4.20
2400	4.27	4.20	4.32	4.25	4.38	4.20
2700	4.15	4.18	4.29	4.01	4.34	4.01
3000	5.02	4.21	4.27	4.10	4.30	4.10
			Quadrature-Phase Component			
1800	4.27	4.42	4.27	4.10	4.27	4.07
2100	4.25	4.21	4.34	4.24	4.25	4.28
2400	4.24	4.27	4.37	4.20	4.37	4.20
2700	5.12	4.16	4.35	4.10	4.34	4.04
3000	5.09	4.26	4.30	4.23	4.27	4.12

Note: Data Set I, Dartmouth.

Moreover, we observe that Tables 9.1–9.5 (which correspond to different radar locations) are more or less consistent in terms of the D_2 values, irrespective of the fact that the data were recorded at different radar locations and using different radar types.

The effect of varying sea state was also studied. The sea states, measured in terms of wave height, change from 0.85 to 2.6 m in Tables 9.1–9.5. We see that the D_2 values are essentially independent of wave height and therefore sea state. Table 9.5 shows the D_2 estimates of raw, three-point smoothed, and FIR filtered clutter data from the experiments at Argentina. Here, it was noted that the raw data were more noisy (low SNR), and because of the influence of noise, the D_2 estimates show higher

Table 9.2   Correlation Dimension of In-Phase, Quadrature Phase, and Amplitude of Clutter Data for Different Ranges

Range (m)	I–Q Corrected		Three-Point Smoothed		FIR Filtered	
	D_{KY}	D_{ML}	D_{KY}	D_{ML}	D_{KY}	D_{ML}
Amplitude						
1590	4.27	4.28	4.18	4.70	4.47	4.33
1620	4.20	4.57	4.20	4.31	4.44	4.78
1650	4.31	4.41	4.10	4.33	4.37	4.51
1680	4.69	4.28	4.45	4.53	4.36	4.69
1710	4.33	4.35	4.47	4.46	4.30	4.25
In-Phase Component						
1590	4.87	4.50	4.43	4.28	4.64	4.61
1620	4.85	4.42	4.20	4.35	4.41	4.21
1650	4.10	4.10	4.21	4.13	4.31	4.05
1680	4.12	4.75	4.17	4.36	4.46	4.17
1710	4.40	4.50	4.18	4.17	4.41	4.11
Quadrature-Phase Component						
1590	4.32	4.75	4.23	4.16	4.45	4.38
1620	4.13	4.85	4.19	4.07	4.46	4.43
1650	4.45	4.91	4.40	4.53	4.44	4.05
1680	4.10	4.27	4.54	4.68	4.35	4.19
1710	4.02	4.23	4.31	4.81	4.36	4.45

Note: Data Set II, Dartmouth.

values for the unfiltered data set. However, the D_2 estimates of the filtered data once again are in the range of 4.1–4.5.

These results show that sea clutter has a strange attractor with a correlation dimension lying between 4.1 and 4.5. This result holds irrespective of radar type, radar location, or sea state. We also observe that the D_2 values for in-phase, quadrature-phase, and amplitude components remain essentially consistent for all data sets. This latter observation indicates that a single radar component should be sufficient for reconstructing the underlying dynamics of the sea clutter, in accordance with the Takens embedding theorem [Mañé, 1981; Packard et al., 1980; Takens, 1981].

Table 9.3   Correlation Dimension of In-Phase, Quadrature Phase, and Amplitude of Clutter Data for Different Ranges

Range (m)	I–Q Corrected		Three-Point Smoothed		FIR Filtered	
	D_{KY}	D_{ML}	D_{KY}	D_{ML}	D_{KY}	D_{ML}
	Amplitude					
2739	4.93	4.74	4.20	4.39	4.27	4.67
2799	4.23	4.17	4.22	4.38	4.28	4.11
2859	4.30	4.38	4.24	4.28	4.29	4.46
2919	4.36	4.24	4.28	4.26	4.33	4.24
2979	4.35	4.26	4.37	4.44	4.35	4.21
	In-Phase Component					
2739	4.62	4.28	4.41	4.20	4.59	4.17
2799	4.71	4.79	4.44	4.14	4.69	4.31
2859	4.72	4.24	4.86	4.60	4.72	4.30
2919	4.68	4.40	4.52	4.55	5.70	4.40
2979	4.97	4.21	4.52	4.36	4.95	4.16
	Quadrature-Phase Component					
2739	4.62	4.54	4.37	4.38	4.60	4.31
2799	4.77	4.50	4.48	4.52	4.26	4.65
2859	4.71	4.26	4.89	4.39	4.72	4.38
2919	5.00	4.58	4.46	4.24	5.70	4.36
2979	4.96	4.42	4.43	4.60	4.96	4.25

Note: Data Set III, Dartmouth.

9.3   LYAPUNOV EXPONENTS

The correlation dimension (D_2) characterizes the distribution of points in the state space, whereas Lyapunov exponents describe the action of the dynamics defining the evolution of the trajectories [Benettin et al., 1980a,b; Eckmann and Ruelle, 1985; Farmer, 1985; Mayer-Kress and Hubler, 1986; Oseledec, 1968; Wolf et al., 1985]. They are the average exponental rates of divergence/convergence of nearby trajectories in phase space [Wolf et al., 1985]. The Lyapunov exponents of a dynamical system are the most important invariants that characterize the system. They are independent of initial conditions on any orbit and are thus properties of the attractor geometry and the dynamics [Benettin et al.,

Table 9.4   Correlation Dimension of In-Phase, Quadrature Phase, and Amplitude of Clutter Data for Different Ranges

Range (m)	I–Q Corrected		Three- Point Smoothed		FIR Filtered	
	D_{KY}	D_{ML}	D_{KY}	D_{ML}	D_{KY}	D_{ML}
Amplitude						
5730	4.86	4.25	4.11	4.17	4.17	4.37
5850	5.78	4.46	4.25	4.19	4.21	4.09
6000	4.80	4.22	4.13	4.10	4.16	4.02
6150	4.93	4.57	4.20	4.31	4.21	4.04
6300	4.99	4.61	4.17	4.21	4.14	4.21
6450	4.57	4.54	4.15	4.26	4.24	4.24
6600	4.31	4.63	4.12	4.17	4.15	4.24
In-Phase Component						
5730	4.46	4.69	4.16	4.32	4.23	4.10
5850	4.81	4.70	4.10	4.32	4.16	4.19
6000	4.55	4.27	4.20	4.29	4.30	4.03
6150	4.45	4.84	4.32	4.23	4.34	4.11
6300	4.70	4.48	4.30	4.14	4.22	4.11
6450	4.39	4.97	4.10	4.32	4.13	4.06
6600	4.20	4.52	4.11	4.19	4.05	4.15
Quadrature-Phase Component						
5730	4.99	4.25	4.10	4.07	4.23	4.34
5850	5.07	4.30	4.13	4.33	4.18	4.17
6000	4.60	4.40	4.26	4.28	4.38	4.10
6150	5.30	4.57	4.52	4.63	4.51	4.10
6300	4.50	4.61	4.21	4.38	4.32	4.14
6450	4.35	4.53	4.10	4.06	4.14	4.14
6600	4.75	4.62	4.10	3.95	4.16	4.08

Note: Data Set IV, Cape Bonavista.

1980a,b; Eckmann and Ruelle, 1985; Farmer, 1985; Mayer-Kress and Hubler, 1986; Oseledec, 1968; Wolf et al., 1985]. For a system to be chaotic, it has to have at least one positive Lyapunov exponent. The magnitude of this exponent defines the time scale on which the system dynamics become unpredictable [Sidorowich, 1992]. Bennettin et al. [1980a,b] showed that the estimation of Lyapunov exponents for systems

Table 9.5   Correlation Dimension of Amplitude of Clutter Data for Different Ranges

Range (m)	Raw		Three-Point Smoothed		FIR Filtered	
	D_{KY}	D_{ML}	D_{KY}	D_{ML}	D_{KY}	D_{ML}
4500	5.63	5.62	4.26	4.41	4.14	4.32
5625	—	7.94	4.96	5.20	4.41	4.79
6900	6.70	6.92	4.38	4.69	4.27	4.37

Note: Data Set V, Argentia.

whose equations of motion are well defined is straightforward [Shimada and Nagashima, 1979]. Unfortunately, their method cannot be applied directly to experimental time series. Wolf et al.'s [1985] method is only suitable for estimating the largest positive Lyapunov exponent of a chaotic process. Moreover, it requires data with high SNR and the attractor dimension should be medium or low; these are difficult conditions to obtain in an experimental set-up like ours.

9.3.1   Estimation of Lyapunov Exponents

There is an abundance of literature available on different methods for the estimation of Lyapunov exponents using an experimental time series [Abarbanel, 1996; Abarbanel and Sushchik, 1993a; Abarbanel et al., 1991a,b, 1992, 1993; Bayna and Tsuda, 1993; Benettin et al., 1978; Benettin et al., 1980a,b; Briggs, 1990; Brown et al., 1991; Bryant et al., 1990; Eckmann and Ruelle, 1985; Eckmann et al., 1986; Farmer, 1985; Geist et al., 1990; Ledrappier, 1981; Mayer-Kress and Hubler, 1986; Oseledec, 1968; Parlitz, 1992; Russell et al., 1980; Sano and Sawada, 1985; Stoop and Meier, 1988; Stoop and Parisi, 1991; von Breman et al., 1997; Wolf et al., 1985; Young, 1982, 1983]. A frequently used method for estimating the Lyapunov exponents of an experimental time series involves recursive QR decomposition applied to the Jacobian of a function $\mathbf{F}[\mathbf{s}(n)]$ that maps points on the orbit into points at t time steps later [Abarbanel, 1996; Abarbanel et al., 1993; Briggs, 1990; Brown et al., 1991; Bryant et al., 1990]:

$$\mathbf{s}(n + t) = \mathbf{F}[\mathbf{s}(n)] \qquad (9.15)$$

Here the map is unknown, and the Jacobian is to be estimated using knowledge of the trajectory points $\mathbf{s}(n)$. The map $\mathbf{F}$ describes nearby trajectories as well as how the distance between these neighbors changes with time. Let $\mathbf{z}^{(r)}(n, 0)$ be the distance between the fiducial orbit and the rth neighbor at $t = 0$. The corresponding distance, time t later, is

$$\mathbf{z}^{(r)}(n, t) = \mathbf{F}[(\mathbf{s}(n) + \mathbf{z}^{(r)}(n, 0)] - \mathbf{F}[\mathbf{s}(n)] \qquad (9.16)$$

A Taylor series expansion of $\mathbf{F}[\cdot]$ contains the Jacobian of the underlying dynamics as the leading term of this expression. The terms in the Jacobian are found by a least-squares minimization of the residuals in the above formula [Briggs, 1990; Brown et al., 1991; Bryant et al., 1990]. The Lyapunov exponents are then calculated as the logarithm of the eigenvalues of products of such Jacobian matrices for a number of steps along the trajectory of the orbit [Abarbanel, 1996; Chaos Software, 1995]. We refer to this algorithm as the BBA algorithm, in recognition of its originators, Brown, Bryant, and Abarbanel [Brown et al., 1991; Bryant et al., 1990]. In an independent publication, Briggs [1990] reported an algorithm similar to that of Brown et al. [1991].

9.3.2 Results of Case Studies

As mentioned previously, the Lyapunov exponents tell us not only about the sensitive dependence of dynamics on the initial conditions of nearly trajectories but also how orbits on the attractor move apart/together under the evolution of the underlying dynamics.

We used the BBA algorithm for estimating the Lyapunov exponents of the sea clutter data [Haykin and Puthusserypady, 1997]. The algorithm gave five or six exponents corresponding to the local embedding dimension used. A data length of 50,000 samples and a minimum of 3000 starting locations were used in each calculation. The Lyapunov exponents are measured in units of *nats per sample*, as explained in Eq. (4.11).

Here, again, as we did in the D_2 analysis, the effects of varying radar parameters and sea states on the Lyapunov spectrum were studied in detail. Tables 9.6–9.10 summarize the Lyapunov spectra for the five sea clutter data sets parameterized in Chapter 3. From these tables, we observe the following:

- For a given radar type and sea state, the Lyapunov exponents are essentially independent of radar range and radar component (in-phase, quadrature phase, or amplitude).

Table 9.6 Lyapunov Exponents[a] of In-Phase, Quadrature Phase, and Amplitude of Clutter Data for Different Ranges[b]

Range (m)	Amplitude			In-Phase Component			Quadrature Phase Component		
	Corrected	Three-Point	FIR	Corrected	Three-Point	FIR	Corrected	Three-Point	FIR
1800	+0.1927	+0.2069	+0.2047	+0.1333	+0.1168	+0.0969	+0.1481	+0.1195	+0.0948
	+0.0966	+0.0947	+0.0876	+0.0558	+0.0387	+0.0244	+0.0539	+0.0401	+0.0312
	+0.0047	−0.0128	+0.0022	−0.0138	−0.0042	−0.0040	−0.0099	−0.0087	−0.0070
	−0.1157	−0.1461	−0.1289	−0.0909	−0.0841	−0.0585	−0.1018	−0.0750	−0.0525
	−0.3023	−0.4825	−0.4753	−0.3602	−0.2954	−0.2383	−0.3443	−0.2799	−0.2438
	−0.6793								
2100	+0.1554	+0.2044	+0.2076	+0.1155	+0.1191	+0.0927	+0.1062	+0.1128	+0.0861
	+0.0766	+0.0835	+0.0973	+0.0476	+0.0450	+0.0310	+0.0508	+0.0429	+0.0246
	−0.0132	−0.0126	+0.0020	−0.0134	−0.0089	−0.0001	−0.0055	−0.0024	−0.0082
	−0.1267	−0.1412	−0.1196	−0.0802	−0.0730	−0.0531	−0.0870	−0.0672	−0.0460
	−0.2763	−0.4712	−0.4323	−0.2757	−0.2678	−0.2039	−0.2663	−0.2533	−0.2219
	−0.6840								
2400	+0.1951	+0.2246	+0.2519	+0.1141	+0.1304	+0.0931	+0.1152	+0.1182	+0.0981
	+0.1084	+0.1137	+0.1146	+0.0450	+0.0496	+0.0260	+0.0402	+0.0444	+0.0294
	+0.0009	−0.0044	+0.0054	−0.0087	−0.0107	−0.0016	−0.0100	−0.0072	−0.0052
	−0.1189	−0.1734	−0.1378	−0.0804	−0.0764	−0.0428	−0.0763	−0.0657	−0.0458
	−0.3082	−0.5295	−0.4885	−0.2737	−0.2907	−0.1978	−0.2932	−0.2458	−0.2043
	−0.7537								

2700	+0.1619	+0.1968	+0.2093	+0.0967	+0.1040	+0.0790	+0.1165	+0.1043	+0.0937
	+0.0668	+0.0948	+0.0993	+0.0361	+0.0417	+0.0296	+0.0539	+0.0421	+0.0271
	−0.0119	−0.0056	+0.0036	−0.0127	−0.0052	−0.0069	+0.0047	+0.0025	−0.0012
	−0.1140	−0.1418	−0.1427	−0.0823	−0.0718	−0.0392	−0.0325	−0.0667	−0.0542
	−0.3045	−0.4605	−0.4457	−0.2648	−0.2378	−0.1849	−0.1055	−0.2335	−0.1955
	−0.7151								
3000	+0.1839	+0.2198	+0.2119	+0.0939	+0.0897	+0.0949	+0.1147	+0.0964	+0.0820
	+0.0822	+0.1028	+0.1080	+0.0378	+0.0331	+0.0231	+0.0536	+0.0357	+0.0307
	−0.0023	−0.0180	+0.0016	+0.0077	−0.0097	−0.0089	+0.0075	−0.0023	−0.0068
	−0.1171	−0.1596	−0.1046	−0.0277	−0.0552	−0.0471	−0.0369	−0.0621	−0.0515
	−0.3037	−0.5063	−0.4378	−0.1057	−0.2149	−0.2120	−0.1091	−0.2246	−0.2039
	−0.7746			−0.3000			−0.3095		

[a] In nats per sample.
[b] Data Set I, Dartmouth.

Table 9.7 Lyapunov Exponents[a] of In-Phase, Quadrature Phase, and Amplitude of Clutter Data for Different Ranges[b]

Range (m)	Amplitude			In-Phase Component			Quadrature Phase Component		
	Corrected	Three-Point	FIR	Corrected	Three-Point	FIR	Corrected	Three-Point	FIR
1590	+0.1998	+0.2325	+0.2693	+0.2730	+0.2692	+0.2929	+0.1888	+0.1728	+0.2422
	+0.0938	+0.1060	+0.1458	+0.1557	+0.1276	+0.1423	+0.0883	+0.0646	+0.1100
	−0.0207	−0.0260	+0.0183	+0.0353	+0.0025	+0.0208	−0.0111	−0.0089	+0.0018
	−0.1658	−0.2017	−0.1603	−0.1248	−0.1537	−0.1292	−0.1445	−0.1237	−0.1193
	−0.3394	−0.6106	−0.5735	−0.3897	−0.5713	−0.5070	−0.3861	−0.4626	−0.5154
	−0.9518			−1.0103			−0.9760		
1620	+0.1976	+0.2334	+0.2804	+0.2890	+0.1855	+0.1983	+0.1671	+0.1666	+0.2158
	+0.0916	+0.1064	+0.1416	+0.1663	+0.0759	+0.0810	+0.0615	+0.0584	+0.1006
	−0.0229	−0.0278	+0.0096	+0.0388	−0.0098	−0.0012	−0.0280	−0.0120	−0.0003
	−0.1889	−0.2062	−0.1630	−0.1444	−0.1482	−0.0978	−0.1502	−0.1262	−0.1132
	−0.4247	−0.6265	−0.6012	−0.4100	−0.5094	−0.4429	−0.3870	−0.4469	−0.4399
	−0.9928			−1.0410			−0.9729		
1650	+0.2151	+0.2068	+0.2482	+0.2242	+0.1822	+0.1434	+0.1314	+0.1231	+0.2286
	+0.0980	+0.0871	+0.1229	+0.1115	+0.0828	+0.0616	+0.0432	+0.0478	+0.1139
	−0.0190	−0.0477	−0.0005	−0.0293	−0.0124	−0.0049	−0.0425	−0.0126	+0.0117
	−0.1600	−0.2112	−0.1666	−0.2570	−0.1460	−0.0908	−0.1660	−0.0849	−0.1295
	−0.4318	−0.6378	−0.5537	−0.7535	−0.5029	−0.3567	−0.3679	−0.1853	−0.5037
	−1.0059						−0.9129	−0.5206	

	1	2	3	4	5	6	7	8	9
1680	+0.2160	+0.1972	+0.2544	+0.1421	+0.1662	+0.1862	+0.1590	+0.1268	+0.1672
	+0.1129	+0.1048	+0.1263	+0.0588	+0.0622	+0.0893	+0.0567	+0.0575	+0.0727
	+0.0021	−0.0093	+0.0007	−0.0248	−0.0166	−0.0037	−0.0308	−0.0072	−0.0092
	−0.1261	−0.1398	−0.1723	−0.1340	−0.1373	−0.1010	−0.1659	−0.0799	−0.1035
	−0.2974	−0.3404	−0.5780	−0.3446	−0.4312	−0.3700	−0.3746	−0.1768	−0.3647
	−1.1682	−0.7734		−0.7468			−0.8638	−0.4654	
1710	+0.2051	+0.1873	+0.2252	+0.1828	+0.1746	+0.2068	+0.1403	+0.0983	+0.2206
	+0.1050	+0.1019	+0.1135	+0.0900	+0.0683	+0.0918	+0.0485	+0.0356	+0.1051
	−0.0073	−0.0072	−0.0125	−0.0066	−0.0186	−0.0015	−0.0300	−0.0132	−0.0012
	−0.1606	−0.1284	−0.1640	−0.1287	−0.1410	−0.1151	−0.1523	−0.0709	−0.1516
	−0.4342	−0.3246	−0.5337	−0.3486	−0.4691	−0.4366	−0.3457	−0.1612	−0.4900
	−1.0249	−0.7560		−0.8480			−0.7810	−0.4143	

[a] In nats per sample.
[b] Data Set II, Dartmouth.

Table 9.8 Lyapunov Exponents[a] of In-Phase, Quadrature Phase, and Amplitude of Clutter Data for Different Ranges[b]

Range (m)	Amplitude			In-Phase Component			Quadrature Phase Component		
	Corrected	Three-Point	FIR	Corrected	Three-Point	FIR	Corrected	Three-Point	FIR
2739	+0.2370	+0.2616	+0.2965	+0.3943	+0.3162	+0.3895	+0.4064	+0.3196	+0.3994
	+0.1386	+0.1437	+0.1468	+0.2262	+0.1603	+0.2378	+0.2253	+0.1590	+0.2299
	+0.0229	−0.0218	−0.0012	+0.0574	+0.0088	+0.0392	+0.0450	−0.0004	+0.0437
	−0.1097	−0.2049	−0.2300	−0.2037	−0.1891	−0.2006	−0.1905	−0.2078	−0.2030
	−0.3125	−0.7274	−0.7851	−0.7684	−0.7264	−0.7896	−0.7847	−0.7227	−0.7778
	−0.8067								
2799	+0.3170	+0.3004	+0.3159	+0.4667	+0.3272	+0.4614	+0.4715	+0.3601	+0.2630
	+0.1658	+0.1495	+0.1743	+0.2811	+0.2008	+0.2743	+0.2787	+0.2149	+0.1184
	−0.0052	−0.0105	−0.0015	+0.0818	+0.0204	+0.0767	+0.1002	+0.0233	−0.0116
	−0.2799	−0.2591	−0.2594	−0.2287	−0.2058	−0.2276	−0.1974	−0.2057	−0.1876
	−0.8664	−0.8109	−0.8024	−0.8431	−0.7753	−0.8465	−0.8527	−0.8117	−0.7120
2859	+0.3421	+0.2841	+0.3240	+0.4256	+0.4902	+0.4102	+0.4288	+0.5096	+0.4186
	+0.1848	+0.1629	+0.1855	+0.2465	+0.2939	+0.2347	+0.2396	+0.3106	+0.2456
	+0.0009	−0.0113	−0.0085	+0.0566	+0.0920	+0.0629	+0.0563	+0.0858	+0.0538
	−0.2770	−0.2412	−0.2556	−0.1800	−0.1849	−0.1729	−0.1891	−0.1927	−0.1885
	−0.8557	−0.8206	−0.8399	−0.7609	−0.8063	−0.7353	−0.7509	−0.8028	−0.7279

2919								
+0.3401	+0.3059	+0.3496	+0.4362	+0.3331	+0.4296	+0.5070	+0.3429	+0.4381
+0.1922	+0.1541	+0.1857	+0.2787	+0.1494	+0.2796	+0.3009	+0.1509	+0.2716
+0.0037	+0.0041	+0.0077	+0.1342	+0.0193	+0.1408	+0.0889	+0.0154	+0.1377
−0.2401	−0.2413	−0.2589	−0.0235	−0.1522	−0.0235	−0.1637	−0.1739	−0.0218
−0.8310	−0.8070	−0.8624	−0.2690	−0.6731	−0.2648	−0.7509	−0.6880	−0.2589
			−0.8157		−0.8026			−0.8144

2979								
+0.3277	+0.3125	+0.3263	+0.4672	+0.3216	+0.4627	+0.4603	+0.3140	+0.4742
+0.1750	+0.1782	+0.1966	+0.3171	+0.1441	+0.2672	+0.2711	+0.1343	+0.2643
−0.0019	−0.0058	+0.0070	+0.1492	+0.0095	+0.0823	+0.0803	+0.0085	+0.0844
−0.2313	−0.2195	−0.2334	+0.0136	−0.1553	−0.1421	−0.1592	−0.1740	−0.1617
−0.7808	−0.7203	−0.8320	−0.2182	−0.6162	−0.7008	−0.6808	−0.6641	−0.6847
			−0.7701					

[a] In nats per sample.
[b] Data set III, Dartmouth.

Table 9.9 Lyapunov Exponents[a] of In-Phase, Quadrature Phase, and Amplitude of Clutter Data for Different Ranges[b]

Range (m)	Amplitude			In-Phase Component			Quadrature Phase Component		
	Corrected	Three-Point	FIR	Corrected	Three-Point	FIR	Corrected	Three-Point	FIR
5730	+0.5052	+0.2504	+0.2389	+0.3736	+0.1955	+0.1943	+0.2827	+0.1903	+0.1941
	+0.3376	+0.1184	+0.1185	+0.2209	+0.0941	+0.0829	+0.1662	+0.0816	+0.0795
	+0.1157	−0.0269	−0.0191	+0.0516	−0.0285	−0.0272	+0.0457	−0.0309	−0.0102
	−0.1997	−0.2599	−0.2214	−0.2492	−0.1730	−0.1376	−0.1264	−0.1852	−0.1441
	−0.8868	−0.7601	−0.6817	−0.8664	−0.5607	−0.4976	−0.3714	−0.5809	−0.5299
	−1.0730						−0.9841		
5850	+0.5686	+0.3420	+0.2902	+0.4722	+0.2186	+0.1922	+0.3354	+0.2108	+0.1927
	+0.4080	+0.1826	+0.1501	+0.3155	+0.1011	+0.0808	+0.2078	+0.1111	+0.0843
	+0.2294	−0.0111	−0.0181	+0.1176	−0.0307	−0.0189	+0.0672	−0.0304	−0.0165
	−0.0128	−0.2874	−0.2488	−0.2071	−0.2403	−0.1593	−0.1230	−0.2049	−0.1518
	−0.3428	−0.8999	−0.8286	−0.8619	−0.6926	−0.5912	−0.4193	−0.6722	−0.5990
	−1.0730						−1.0303		
6000	+0.5032	+0.2591	+0.2278	+0.4204	+0.2713	+0.2519	+0.4221	+0.2625	+0.3060
	+0.3209	+0.1215	+0.1157	+0.2439	+0.1305	+0.1298	+0.2516	+0.1368	+0.1682
	+0.1166	−0.0239	−0.0179	+0.0712	−0.0257	−0.0111	+0.0579	−0.0056	−0.0032
	−0.2283	−0.2645	−0.2200	−0.2562	−0.2371	−0.1932	−0.2215	−0.2115	−0.2073
	−0.8903	−0.7352	−0.6528	−0.8767	−0.7275	−0.5884	−0.8562	−0.6925	−0.6825

140

6150	+0.2856	+0.2593	+0.2603	+0.3696	+0.2712	+0.2384	+0.3812	+0.3786	+0.3282
	+0.1700	+0.1244	+0.1351	+0.2039	+0.1493	+0.1164	+0.2501	+0.2149	+0.1608
	+0.0445	-0.0219	-0.0162	+0.0346	-0.0065	+0.0021	+0.1023	+0.0429	+0.0227
	-0.1343	-0.2275	-0.2395	-0.2295	-0.1924	-0.1596	-0.0749	-0.2310	-0.1821
	-0.3930	-0.7168	-0.6800	-0.8486	-0.6921	-0.5758	-0.3669	-0.7797	-0.6472
	-0.9669						-0.9662		
6300	+0.5427	+0.2676	+0.2369	+0.4657	+0.2532	+0.1956	+0.4263	+0.2658	+0.2594
	+0.3659	+0.1382	+0.1151	+0.2783	+0.1441	+0.0765	+0.2543	+0.1251	+0.1391
	+0.1458	-0.0143	-0.0285	+0.0789	-0.0056	-0.0175	+0.0485	-0.0134	-0.0058
	-0.1774	-0.2573	-0.2250	-0.2195	-0.1914	-0.1362	-0.2649	-0.2308	-0.1829
	-0.8889	-0.8092	-0.7069	-0.8607	-0.6604	-0.5476	-0.9448	-0.7071	-0.6651
6450	+0.4192	+0.2550	+0.2644	+0.3566	+0.1670	+0.1377	+0.3222	+0.1610	+0.1401
	+0.2673	+0.1347	+0.1376	+0.2006	+0.0704	+0.0386	+0.1784	+0.0660	+0.0566
	+0.0507	-0.0209	-0.0152	+0.0035	-0.0296	-0.0099	+0.0118	-0.0324	-0.0186
	-0.2269	-0.2588	-0.2216	-0.2336	-0.1702	-0.1126	-0.2337	-0.1505	-0.1185
	-0.8992	-0.7493	-0.6773	-0.8522	-0.5251	-0.4304	-0.7957	-0.5450	-0.4426
6600	+0.3368	+0.2241	+0.2323	+0.2718	+0.1660	+0.1143	+0.2631	+0.1791	+0.1847
	+0.1996	+0.1058	+0.1082	+0.1369	+0.0695	+0.0406	+0.1326	+0.0696	+0.0650
	+0.0240	-0.0165	-0.0182	-0.0331	-0.0230	-0.0206	-0.0194	-0.0301	-0.0184
	-0.2856	-0.2305	-0.2265	-0.2281	-0.1565	-0.1127	-0.2396	-0.1680	-10.466
	-0.8960	-0.6690	-0.6473	-0.7430	-0.4979	-0.3995	-0.7670	-0.5300	-0.5456

[a] In nats per sample.
[b] Data Set IV, Cape Bonavista.

141

**Table 9.10   Lyapunov Exponents[a] of Clutter
Amplitude for Different Ranges[b]**

Range (m)	Raw	Three-Point	FIR
4500	+0.5123	+0.3175	+0.2435
	+0.3595	+0.1695	+0.1128
	+0.1759	−0.0060	−0.0295
	−0.0305	−0.2580	−0.2305
	−0.3464	−0.8652	−0.7008
	−1.0701		
5625	+0.7668	+0.5332	+0.3543
	+0.6627	+0.3622	+0.1452
	+0.5619	+0.1402	−0.0052
	+0.4483	−0.1780	−0.1921
	+0.3455	−0.8994	−0.7365
	+0.2149		
	+0.0541		
	−0.1719		
	−0.4825		
	−1.2132		
6900	+0.5267	+0.3349	+0.2848
	+0.3982	+0.1980	+0.1485
	+0.2608	+0.0295	+0.0032
	+0.1118	−0.2332	−0.2229
	−0.1095	−0.8751	−0.7926
	−0.4201		
	−1.0938		

[a] In nats per sample.
[b] Data Set V, Argentia.

- For a given radar type, the Lyapunov spectrum exhibits significant dependence on sea state (see Tables 9.7 and 9.8 pertaining to the same radar system and radar site but different wave heights). In particular, the magnitudes of the positive Lyapunov exponents increase with increasing wave height, which implies less predictability as the sea state increases. This form of behavior is intuitively satisfying.

- Most importantly, in every case, the Lyapunov spectrum exhibits the same structure: two positive exponents, the third exponent close to zero, followed by two or three negative exponents (depending on the embedding dimension), with the sum of all Lyapunov exponents being negative.

Table 9.10 corresponds to the sea clutter data set from Argentia, Newfoundland. Because of the presence of noise in this case, the raw data show more than two positive Lyapunov exponents, whereas the filtered data behave in a similar way to that described above (with two positive exponents, the third one approximately zero, followed by negative exponents). For example, we observe that the Lyapunov spectrum of the raw data corresponding to range 5625 m has 10 exponents, with 7 of them being positive and with a positive sum, and the BBA algorithm did not even compute the Kaplan–Yorke dimension (see Table 9.5). The reason for this nonchaotic behavior is the presence of a significant amount of noise in the sea clutter data.

Figure 9.2 illustrates the Lyapunov exponents for a typical FIR filtered sea clutter data set collected during the experiment at Argentia, Newfoundland. The Lyapunov exponents are shown plotted as a function of L, denoting the number of times for which the linear terms in the Jacobian matrices used in the BBA algorithm are multiplied together. The exponents have been averaged over 3000 different starting locations on the attractor. We observe from this figure that there are two positive exponents and two negative exponents, with the third exponent approaching zero. For this computation, we used an order of fit (for the polynomial) equal to 2.

The important conclusions drawn from the Lyapunov spectrum analysis of sea clutter (computed using the BBA algorithm) may be summarized as follows:

- In all the cases (except for the noticeably noisy ones), the BBA algorithm invariably produced two positive exponents, which is further evidence of chaotic dynamics of sea clutter.
- The third exponent goes to zero consistently, indicating that sea clutter can be modeled as a set of coupled nonlinear differential equations.
- The sum of all exponents is negative, indicating a dissipative dynamical and hence a physically realizable system.

Figures 9.3–9.5 show the average (over different ranges) of the largest Lyapunov exponent versus the wave height plots for the amplitude, in-

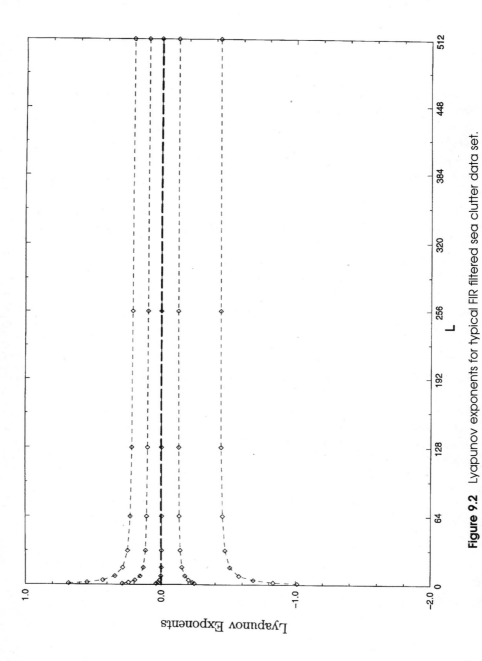

Figure 9.2 Lyapunov exponents for typical FIR filtered sea clutter data set.

144

phase, and quadrature-phase components of the clutter data from the Dartmouth database. Figures 9.3*a–c* are plots of the variation of the average of the largest Lyapunov exponent to the wave height for the unfiltered (I–Q corrected), three-point smoothed, and FIR filtered amplitude components of the clutter data. It can be seen that as the sea state becomes more and more rough, the largest Lyapunov exponent increases.

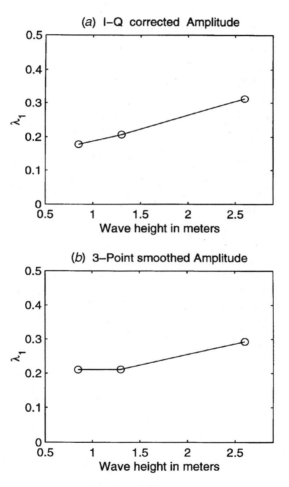

Figure 9.3 Average of largest Lyapunov exponent vs. wave height for amplitude component: (*a*) I–Q corrected amplitude component; (*b*) I–Q corrected three-point smoothed amplitude.

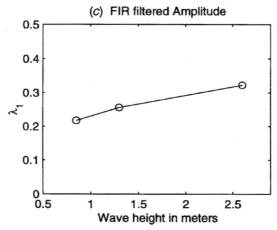

Figure 9.3 Average of largest Lyapunov exponent vs. wave height for amplitude component: (c) I–Q corrected FIR filtered amplitude.

Intuitively, this is to be expected because, as the sea state becomes more and more rough, the wave height increases and the dynamics of the system becomes less predictable. This is shown up in the increase in the magnitude of the largest Lyapunov exponent. Similar behavior has been observed in the case of in-phase and quadrature-phase components of the sea clutter data and is shown in Figs. 9.4 and 9.5, respectively.

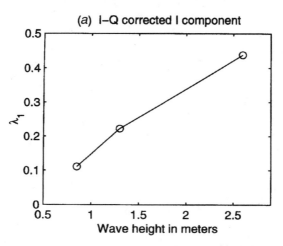

Figure 9.4 Average of largest Lyapunov exponent vs. wave height for in-phase component: (a) I–Q corrected I-phase component.

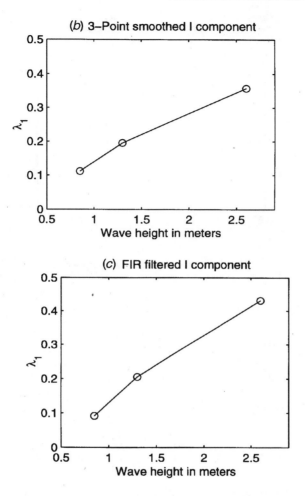

Figure 9.4 Average of largest Lyapunov exponent vs. wave height for in-phase component: (*b*) I–Q corrected three-point smoothed I-phase; (*c*) I–Q corrected FIR filtered I-phase.

9.4 HORIZON OF PREDICTABILITY (HP)

Short-term predictability is an important characteristic of a chaotic process. This property is attributed to the presence of positive Lyapunov exponents, which try to pull apart trajectories that are close to each other in the phase space. The HP (where the trajectories are close to each other)

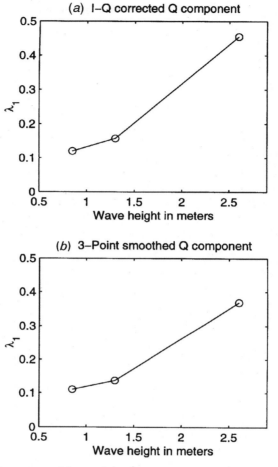

Figure 9.5 Average of largest Lyapunov exponent vs. wave height for quadrature-phase component: (a) I–Q corrected Q-phase component; (b) I–Q corrected three-point smoothed Q-phase.

is influenced by the magnitude of the largest Lyapunov exponent. The higher the magnitude of the largest exponent, the lesser will be the number of samples that we can predict, and vice versa. This means that there exists an inverse relationship between the HP and the largest positive Lyapunov exponent.

The BBA algorithm used for the calculation of the Lyapunov spectrum of the sea clutter data gives the HP as a by-product [Brown et al., 1991]. This is estimated as the time required, on average, for trajectories that are within 1% (of root mean square attractor size, i.e., initial separation of the

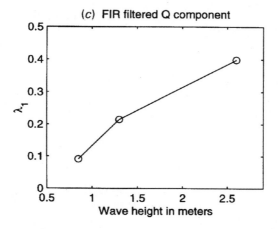

Figure 9.5 Average of largest Lyapunov exponent vs. wave height for quadrature-phase component: (c) I–Q corrected FIR Filtered Q-phase.

Table 9.11 Horizon of Predictability[a] of In-Phase, Quadrature Phase, and Amplitude of Clutter Data for Different Ranges (Dartmouth)

Range	I–Q Corrected	Three-Point Smoothed	FIR Filtered
		Amplitude	
1800	20.30	18.91	19.11
2100	25.17	19.14	18.84
2400	20.05	17.42	15.53
2700	24.16	19.88	18.69
3000	21.27	17.80	18.46
		In-Phase Component	
1800	29.34	33.51	40.38
2100	33.88	32.85	42.20
2400	34.28	30.00	42.03
2700	40.44	37.63	49.54
3000	41.65	43.61	41.23
		Quadrature-Phase Component	
1800	26.41	32.74	41.29
2100	36.84	34.69	45.44
2400	33.96	33.10	39.98
2700	33.57	37.50	41.76
3000	34.12	40.60	47.68

[a] Number of sampling periods.

two trajectories) to separate to 50%. From the definition of Lyapunov exponents, we can write this as

$$\frac{0.50}{0.01} = e^{\lambda_1 t} \tag{9.17}$$

where λ_1 is the largest positive Lyapunov exponent. From the above relationship, the HP t can be estimated as

$$t = \frac{\ln(0.50/0.01)}{\lambda_1} \tag{9.18}$$

The HP over different ranges of sea clutter data (amplitude, in-phase, and quadrature phase) for the Dartmouth database are tabulated in Tables 9.11–9.13, where the HP is expressed in terms of the number of sampling periods. It can be seen that as the sea state increases, the magnitude of the

Table 9.12 Horizon of Predictability[a] of In-Phase, Quadrature Phase, and Amplitude of Clutter Data for Different Ranges (Dartmouth)

Range	I–Q Corrected	Three-Point Smoothed	FIR Filtered
		Amplitude	
1590	19.58	16.82	14.53
1620	19.80	16.77	13.95
1650	18.19	18.92	15.76
1680	18.11	19.83	15.38
1710	19.08	20.88	17.37
		In-Phase Component	
1590	14.33	14.53	13.36
1620	13.54	21.09	19.73
1650	17.45	21.47	27.28
1680	27.52	23.54	21.01
1710	21.41	22.40	18.92
		Quadrature-Phase Component	
1590	20.72	22.64	16.15
1620	23.41	23.48	18.13
1650	29.77	31.77	17.12
1680	24.60	30.85	23.40
1710	27.89	39.79	17.74

[a] Number of sampling periods.

Table 9.13 Horizon of Predictability[a] of In-Phase, Quadrature Phase and Amplitude of Clutter Data for Different Ranges (Dartmouth)

Range	I–Q Corrected	Three-Point Smoothed	FIR Filtered
		Amplitude	
2739	16.15	14.95	13.19
2799	12.34	13.02	12.38
2859	11.44	13.77	12.08
2919	11.50	12.79	11.19
2979	11.94	12.52	11.99
		In-Phase Component	
2739	9.92	12.37	10.04
2799	8.38	11.96	8.48
2859	9.19	7.98	9.54
2919	8.97	11.75	9.11
2979	8.37	12.16	8.46
		Quadrature Component	
2739	9.63	12.24	9.76
2799	8.30	10.86	14.87
2859	9.12	7.68	9.35
2919	7.72	12.04	8.93
2979	8.50	12.46	8.25

[a] Number of sampling periods.

largest Lyapunov exponent increases and the HP decreases. The average of the HP (over different ranges) is plotted against the sea state and is shown in Figs. 9.6–9.8, which confirm the observations we made earlier: As the wave height increases, the HP decreases.

9.5 LYAPUNOV DIMENSION

The Kaplan–Yorke dimension is estimated once we have the Lyapunov spectrum estimated using the formula [see Eq. (4.24)]

$$D_{KY} = K + \frac{\sum_{a=1}^{K} \lambda_a}{|\lambda_{K+1}|} \qquad (9.19)$$

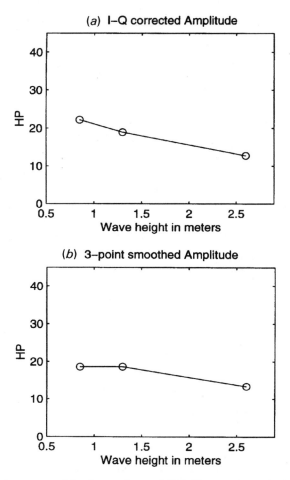

Figure 9.6 Average of horizon of predictability vs. wave height for amplitude component: (*a*) I–Q corrected amplitude component: (*b*) I–Q corrected three-point smoothed amplitude.

where, $\lambda_1 \geq \lambda_2 \geq \cdots \geq \lambda_K \geq 0$ and $\sum_{a=1}^{K} \lambda_a > 0$, $\sum_{a=1}^{K+1} \lambda_a < 0$. Clearly, $K < n$.

9.5.1 Results of Case Studies

The Lyapunov dimension or the Kaplan–Yorke dimension for sea clutter was estimated from the Lyapunov exponents using Eq. (9.19). As

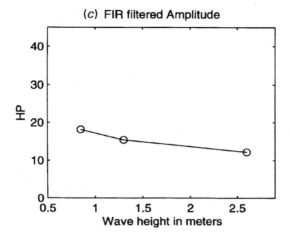

Figure 9.6 Average of horizon of predictability vs. wave height for amplitude component: (c) I–Q corrected FIR filtered amplitude.

mentioned in Chapter 3, this is conjectured to be the same as the fractal dimension estimator, called the information dimension. The results presented in Tables 9.1–9.5 show that it did indeed agree very well with the estimated D_2. This is yet another remarkable evidence for the chaotic dynamics of sea clutter. However, due to the unavoidable presence of

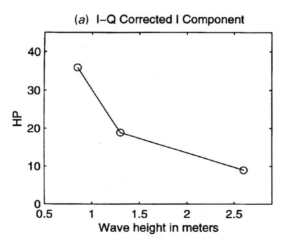

Figure 9.7 Average of horizon of predictability vs. wave height for in-phase component: (a) I–Q corrrected I-phase component.

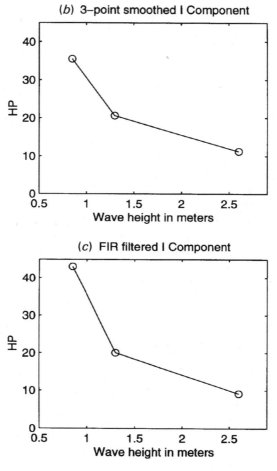

Figure 9.7 Average of horizon of predictability vs. wave height for in-phase component: (*b*) I–Q corrected three-point smoothed I-phase; (*c*) I–Q corrected FIR filtered I-phase.

noise (even after filtering), the equality of D_2 and D_{KY} cannot always be guaranteed. Furthermore, the different algorithms used for the estimation of D_2 and the Lyapunov spectrum (from which D_{KY} is estimated) may not behave in identical ways with respect to the presence of noise in the preprocessed clutter data. It is therefore not surprising to see that D_2 is sometimes slightly greater than D_{KY}, and some other times the other way

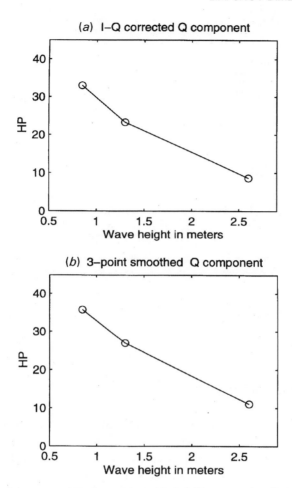

Figure 9.8 Average of horizon of predictability vs. wave height for quadrature-phase component: (*a*) I–Q corrected Q-phase component; (*b*) I–Q corrected three-point smoothed Q-phase.

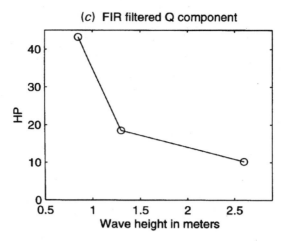

Figure 9.8   Average of horizon of predictability vs. wave height for quadrature-phase component: (*c*) I–Q corrected FIR filtered Q-phase.

round.[2] However, the important point to observe here is that D_2 and D_{KY} are consistently very close to each other.

9.6   KOLMOGOROV ENTROPY

The Kolmogorov entropy, also known as metric entropy or Kolmogorov–Sinai entropy, is the most important measure by which a chaotic motion in an arbitrary dimensional phase space can be characterized [Kolmogorov, 1958]. As discussed in Chapter 4, the Kolmogorov entropy of an attractor can be considered as a measure for the rate of information loss along the attractor or as a measure for the degree of predictability of points along the attractor given an arbitrary initial point. The above definition is made clear in the following sense. Due to the sensitivity to initial conditions in chaotic systems, nearby orbits diverge. If we can only distinguish orbit locations in phase space to within some given accuracy, then the initial conditions for two orbits may appear to be the same. As their orbits evolve forward in time, however, they will eventually move far enough apart that

[2] According to results on the Hénon map, Kaplan and Yorke map, and Zaslavskii map reported by Russell et al. [1980], a similar behavior concerning the estimated values of D_2 and D_{KY} could be observed.

they may be distinguished as different. Alternatively, as an orbit is iterated, by observing its location with the given accuracy that we have, initially insignificant digits in the specification of the initial condition will eventually make themselves felt. Thus, assuming that we can calculate exactly and that we know the dynamical equations defining an orbit, if we view that orbit with limited precision, we can, in principle, use our observation to obtain more and more information about the initial unresolved digits specifying the initial condition. It is in this sense that we say that a chaotic orbit creates information.

In general, a positive, finite entropy is considered as the conclusive proof that the time series and its underlying dynamic process are chaotic.

Examples from statistical mechanics show that disorder is essentially a concept from information theory [Schuster, 1988]. It is therefore not too surprising that the Kolmogorov entropy, which measures "how chaotic a dynamical system is," can also be defined by Shannon's formula in such a way that the Kolmogorov entropy becomes proportional to the rate at which information about the state of the dynamical system is lost in the course of time.

9.6.1 How to Estimate the Kolmogorov Entropy

According to the definitions of the order-2 Kolmogorov entropy as suggested by Takens [1983a] and by Grassberger and Procaccia [1983; Grassberger, 1987], the entropy can be calculated from the average time required for two orbits to diverge that are initially very close to each other. More precisely, the entropy can be calculated from the average time t_0 that is needed for two points on the attractor, which are initially within a specified maximum distance l_0, to separate until the distance between these points has become larger than l_0.

According to Takens [1983a] and Grassberger and Procaccia [1983b], the separation of nearby points on different orbits is assumed to be exponential, and the time interval t_0 required for two initially nearby points to separate by a distance larger than l_0 will be exponentially distributed according to

$$C(t_0) \approx e^{-Kt_0} \tag{9.20}$$

where K is the Kolmogorov entropy.

The points in an experimental time series are measured at discrete, constant time intervals (uniform sampling) with a time step T between two sampled data points, where T is the reciprocal of the sampling frequency (f_s). Consequently, for all practical purposes, $C(t_0)$ should be transformed into a discrete distribution function that is defined as

$$C(b) = e^{-KbT} \qquad (9.21)$$

with $b = 1, 2, \ldots$.

This cumulative distribution function describes the exponential decrease as a function of b. The variable b is the number of sequential pairs of points on the attractor, given an initial pair of independent points within a distance l_0, in which the interpoint distance is for the first time bigger than the specified interpoint distance l_0. In other words, b is obtained from the number of times that

$$\|\mathbf{s}(i + b - 1) - \mathbf{s}(j + b - 1)\| \leq l_0 \qquad (9.22)$$

with $b = 1, 2, 3, \ldots$, provided that

$$\|\mathbf{s}(i) - \mathbf{s}(j)\| \leq l_0 \qquad (9.23)$$

while

$$\|\mathbf{s}(i + b) - \mathbf{s}(j + b)\| > l_0 \qquad (9.24)$$

where $\mathbf{s}(i)$ and $\mathbf{s}(j)$ are points on the reconstructed attractor represented by their state vectors. The maximum distance l_0 can be based on any norm definition. This can be the Euclidean norm or the maximum norm. There is a certain advantage in using the maximum norm over the Euclidean norm, because b is readily obtained by counting the number of times that the absolute difference between sequential pairs of data points in the time series is smaller than l_0, given an initial pair of independent data points $(s(i), s(j))$, according to

$$|s(i + m - 1 + b - 1) - s(j + m - 1 + b - 1)| \leq l_0 \qquad (9.25)$$

for $b = 1, 2, 3, \ldots$ and provided that

$$\max|s(i + k) - s(j + k)| \leq l_0 \qquad (9.26)$$

with $0 \leq k \leq m - 1$, while

$$|s(i + m - 1 + b) - s(j + m - 1 + b)| > l_0 \qquad (9.27)$$

This makes the algorithm fast to compute. We may thus estimate b very quickly and directly from the time series. The rate of the exponential decrease with b is thus measured by the invariant entropy K.

9.6.2   Maximum-Likelihood Estimation of the Entropy

For the estimation of Kolmogorov entropy from an observed time series, several methods have been reported in the literature using the correlation integral. The maximum-likelihood method is preferred over all other methods for its reliable estimates and, moreover, the estimation is to be performed only once. This method is based on the principle that treats the correlation integral as a probability distribution. A maximum-likelihood estimator for the Kolmogorov entropy was derived by Olofsen et al. [1992] that is based on the method proposed by Grassberger and Procaccia [1983b] of calculating the entropy from the quotient of two correlation integrals at large embeddings. Schouten et al. [1994b] came up with a much more elegant maximum-likelihood method that is simple and clear. In this section, we describe the latter method in detail.

For mathematical simplicity, introduce $k = KT$ in Eq. (9.21). The distribution function $C(b)$ can be rewritten as

$$C(b) = e^{-kb} \tag{9.28}$$

The probability of finding a distance bigger than l_0 after exactly b inter-point distances is

$$
\begin{aligned}
p(b) &= C(b-1) \\
&= e^{-k(b-1)} - e^{-kb} \\
&= (e^k - 1)e^{-kb}
\end{aligned}
\tag{9.29}
$$

This probability density function is known as the *geometric* probability density function [Clarke and Disney, 1970]. It satisfies the property

$$\sum_{b=1}^{\infty} p(b) = (e^k - 1) \sum_{b=1}^{\infty} e^{-kb} = 1 \tag{9.30}$$

using the fact that

$$a + a^2 + a^3 + \cdots = \frac{a}{1-a} \tag{9.31}$$

assuming $|a| < 1$.

Using the probability distribution of b, an expression for the entropy based on the maximum-likelihood method can be derived as described here. The probability of finding exactly the sample $(b_1, b_2, \ldots, b_M)$, depending on k, from a random drawing of M pairs of independent points on the attractor is

$$p_k = p(b_1, b_2, \ldots, b_M; k)$$
$$= \prod_{i=1}^{M} p(b_i)$$
$$= (e^k - 1)^M \exp\left(-k \sum_{i=1}^{M} b_i\right) \qquad (9.32)$$

By taking the logarithm of both sides of this equation, we get the log-likelihood function, denoted by $L(k)$; that is,

$$L(k) = \ln[p(b_1, b_2, \ldots, b_M; k)]$$
$$= M \ln(e^k - 1) - k \sum_{i=1}^{M} b_i. \qquad (9.33)$$

Finding the maximum of this function $L(k)$ is equivalent to finding the value of k (or entropy K) that gives the largest probability of finding the sample $(b_1, b_2, \ldots, b_M)$. The maximum of $L(k)$ can be found by differentiating $L(k)$ with respect to k and then equating the result to zero, which yields the equation $\partial L(k)/\partial k$

$$\frac{M}{1 - e^{-k}} = \sum_{i=1}^{M} b_i \qquad (9.34)$$

From Eq. (9.34), the maximum-likelihood estimate of the entropy (k_{ML}) is derived as

$$k_{\mathrm{ML}} = -\ln\left(1 - \frac{1}{\bar{b}}\right) \qquad (9.35)$$

Correspondingly, we may write

$$K_{\mathrm{ML}} = -\frac{1}{T}\ln\left(1 - \frac{1}{\bar{b}}\right) \qquad (9.36)$$

where T is the sampling period. The quantity $\bar{b}$ is the average value of the b's in the sample $(b_1, b_2, \ldots, b_M)$ with sample size M; it is defined as

$$\bar{b} = \frac{1}{M} \sum_{i=1}^{M} b_i \qquad (9.37)$$

It can be proved that the value of K obtained from the above expression is indeed maximum as the second derivative of that quantity is clearly less than zero [Schouten et al., 1994]. Also, it is clear that the estimate of K is only a function of the sample average of b.

9.6.3   Results of Case Studies

In this section, we discuss the results on the estimation of Kolmogorov entropy of sea clutter data for various sea state conditions and radar parameters. Specifically, the maximum-likelihood estimate of the entropy (KE_{ML}) and the sum of the positive Lyapunov exponents (KE_{LE}) are compared. In each instance we find that they both match each other fairly well.

Table 9.14   Kolmogorov Entropy[a] of In-Phase, Quadrature-Phase, and Amplitude Components of Clutter Data for Different Ranges[b]

Range (m)	I–Q Corrected		Three-point Smoothed		FIR Filtered	
	KE_{LE}	KE_{ML}	KE_{LE}	KE_{ML}	KE_{LE}	KE_{ML}
	Amplitude Component					
1800	0.2940	0.2648	0.3016	0.2726	0.2975	0.2403
2100	0.2320	0.2367	0.2879	0.2759	0.3049	0.2654
2400	0.3044	0.2853	0.3383	0.2616	0.3019	0.2464
2700	0.2287	0.2493	0.2916	0.2377	0.3086	0.2464
3000	0.2661	0.2797	0.3226	0.2971	0.3199	0.2913
	In-Phase Component					
1800	0.1891	0.2101	0.1555	0.1855	0.1213	0.1768
2100	0.1631	0.1654	0.1641	0.1609	0.1237	0.1704
2400	0.1591	0.1560	0.1800	0.1739	0.1191	0.1508
2700	0.1328	0.1623	0.1457	0.1565	0.1086	0.1348
3000	0.1394	0.1551	0.1228	0.1493	0.1180	0.1336
	Quadrature-Phase Component					
1800	0.2020	0.2072	0.1596	0.1826	0.1260	0.1551
2100	0.1570	0.1569	0.1557	0.1725	0.1107	0.1348
2400	0.1554	0.1738	0.1626	0.1676	0.1275	0.1676
2700	0.1751	0.1880	0.1489	0.1594	0.1208	0.1591
3000	0.1758	0.1739	0.1321	0.1352	0.1127	0.1319

[a] In nats per sample time step.
[b] Data Set I, Dartmouth.

We used STB_2 algorithm to estimate the Kolmogorov entropy of different sea clutter data. A data length of 50,000 samples was used in each calculation. The entropy is measured in units of nats per sample (as the sampling period was normalized to the pulse repetition frequency of the radar). Tables 9.14–9.18 summarize the Kolmogorov entropy (along with its estimate from the Lyapunov spectrum as the sum of positive Lyapunov exponents) for the five sea clutter data sets parameterized in Chapter 3. From these tables, we observe the following:

1. For a given radar type, the Kolmogorov entropy exhibits significant dependence on sea state. In particular, the magnitude of the entropy increases with the sea state, which implies less predictability in

Table 9.15 Kolmogorov Entropy[a] of In-Phase, Quadrature-Phase, and Amplitude Components of Clutter Data for Different Ranges[b]

Range (m)	I–Q Corrected		Three-Point Smoothed		FIR Filtered	
	KE_{LE}	KE_{ML}	KE_{LE}	KE_{ML}	KE_{LE}	KE_{ML}
	Amplitude Component					
1590	0.2936	0.2821	0.3385	0.3382	0.4334	0.4076
1620	0.2892	0.2945	0.3398	0.3766	0.4316	0.4103
1650	0.3131	0.3151	0.2939	0.3033	0.3711	0.3645
1680	0.3310	0.3084	0.3020	0.3084	0.3814	0.3631
1710	0.3101	0.3443	0.2892	0.2777	0.3387	0.3672
	In-Phase Component					
1590	0.4640	0.4924	0.3993	0.3973	0.2560	0.2979
1620	0.4941	0.5158	0.2614	0.2678	0.2793	0.2922
1650	0.3357	0.3400	0.2650	0.2864	0.2050	0.2141
1680	0.2009	0.2129	0.2284	0.2288	0.2755	0.2702
1710	0.2728	0.2671	0.2429	0.2655	0.2986	0.3108
	Quadrature-Phase Component					
1590	0.2771	0.2978	0.2374	0.2303	0.3540	0.3679
1620	0.2286	0.2216	0.2250	0.2320	0.3164	0.3100
1650	0.1746	0.2010	0.1709	0.1916	0.3542	0.3452
1680	0.2157	0.2137	0.1843	0.1896	0.2392	0.2459
1710	0.1888	0.2143	0.1339	0.1375	0.3257	0.3266

[a] In nats per sample time step.
[b] Data Set II, Dartmouth.

Table 9.16   Kolmogorov Entropy[a] of In-Phase, Quadrature-Phase, and Amplitude Components of Clutter Data for Different Ranges[b]

Range (m)	I–Q Corrected		Three-Point Smoothed		FIR Filtered	
	KE_{LE}	KE_{ML}	KE_{LE}	KE_{ML}	KE_{LE}	KE_{ML}
	Amplitude Component					
2739	0.3985	0.3992	0.4053	0.4125	0.4433	0.4298
2799	0.4828	0.4926	0.4499	0.4727	0.4902	0.4980
2859	0.5278	0.5061	0.4470	0.4421	0.5095	0.5171
2919	0.5360	0.5063	0.4641	0.4677	0.4430	0.4509
2979	0.5027	0.5385	0.4907	0.4847	0.5299	0.5245
	In-Phase Component					
2739	0.6779	0.6628	0.4853	0.4733	0.6665	0.6568
2799	0.8296	0.7224	0.5484	0.5686	0.8124	0.7539
2859	0.7287	0.7437	0.8761	0.8897	0.7078	0.7092
2919	0.8491	0.8387	0.5018	0.5084	0.8500	0.8300
2979	0.9471	0.9501	0.4752	0.4676	0.8122	0.8018
	Quadrature-Phase Component					
2739	0.6767	0.6607	0.4786	0.4857	0.6730	0.6513
2799	0.8504	0.8293	0.5983	0.5852	0.5814	0.5997
2859	0.7247	0.7249	0.9060	0.9293	0.7180	0.7288
2919	0.8968	0.8644	0.5092	0.5176	0.8474	0.8191
2979	0.8117	0.8422	0.4568	0.4576	0.8229	0.8207

[a] In nats per sample time step.
[b] Data Set III, Dartmouth.

terms of large information lost per sample. This was also evident from the Lyapunov spectrum.

2. In almost all cases, the maximum-likelihood estimate of the entropy and its counterpart estimate from the Lyapunov spectrum estimate are close to each other.

3. In cases where the data set is noisy, the maximum-likelihood estimate of the entropy is less than that estimated from the Lyapunov spectrum (Tables 9.17 and 9.18), which may be attributed to the fact that the STB_2 algorithm is robust with respect to the presence of noise in the observed time series. This is particularly evident from the entropy estimates of the unfiltered amplitude, in-phase, and quadrature-phase components.

Table 9.17   Kolmogorov Entropy[a] of In-Phase, Quadrature-Phase, and Amplitude Components of Clutter Data for Different Ranges[b]

Range (m)	I–Q Corrected		Three-Point Smoothed		FIR Filtered	
	KE_{LE}	KE_{ML}	KE_{LE}	KE_{ML}	KE_{LE}	KE_{ML}
			Amplitude Component			
5730	0.9585	0.6554	0.3668	0.3747	0.3574	0.3653
5850	1.2060	0.7327	0.5246	0.5280	0.4403	0.4354
6000	0.9407	0.7760	0.3806	0.3885	0.3435	0.3070
6150	0.5001	0.4877	0.3837	0.3670	0.3954	0.3899
6300	1.0544	0.8125	0.4058	0.4020	0.3520	0.3323
6450	0.7372	0.6251	0.3897	0.4094	0.4020	0.4124
6600	0.5604	0.5030	0.3299	0.3263	0.3405	0.3460
			In-Phase Component			
5730	0.6461	0.5826	0.2896	0.2860	0.2772	0.2859
5850	0.9053	0.8666	0.3197	0.3230	0.2730	0.2650
6000	0.7355	0.6268	0.4018	0.3997	0.3817	0.4024
6150	0.6081	0.5916	0.4205	0.4319	0.3569	0.3206
6300	0.8229	0.7096	0.3973	0.4043	0.2721	0.2843
6450	0.5607	0.4888	0.2374	0.2413	0.1763	0.2043
6600	0.4087	0.4178	0.2355	0.2351	0.1549	0.1603
			Quadrature-Phase Component			
5730	0.4946	0.4194	0.2719	0.2849	0.2736	0.2905
5850	0.6104	0.5773	0.3219	0.3112	0.2770	0.2673
6000	0.7316	0.5883	0.3993	0.3909	0.4742	0.4767
6150	0.7336	0.6607	0.6364	0.6102	0.5117	0.5072
6300	0.7291	0.5991	0.3909	0.3846	0.3985	0.3745
6450	0.5124	0.4656	0.2270	0.2428	0.1967	0.2000
6600	0.3957	0.3989	0.2487	0.2579	0.2497	0.2545

[a] In nats per sample time step.
[b] Data Set IV, Cape Bonavista.

Table 9.18   Kolmogorov Entropy[a] of Amplitude Component of Clutter Data for Different Ranges[b]

Range (m)	I–Q Corrected		Three-Point Smoothed		FIR Filtered	
	KE_{LE}	KE_{ML}	KE_{LE}	KE_{ML}	KE_{LE}	KE_{ML}
	Amplitude Component					
4500	1.0477	0.9049	0.4870	0.4887	0.3563	0.4042
5625	3.0542	0.8431	1.0356	0.7213	0.4995	0.5149
6900	1.2975	0.8710	0.5624	0.5600	0.4365	0.4048

[a] In nats per sample time step.
[b] Data Set V, Argentia.

10

CHAOTIC STUDY OF SIMULATED SEA CLUTTER DATA

10.1 INTRODUCTION

The results of estimated chaotic invariants of sea clutter presented in the previous chapter were all based on real-life data obtained under different environmental conditions and using different radar configurations. In this chapter, we present some corresponding results obtained using simulated sea clutter data. The primary motivation for this presentation is to demonstrate that care has to be exercised when simulated sea clutter data are proposed as a substitute for real-life sea clutter data. The simulation is based on physical considerations of sea clutter as described elsewhere [Haykin et al., 1994].

10.2 OVERVIEW OF THE SIMULATION MODEL

In simulating sea clutter dynamics, several processes must be modeled. These are:

- Sea surface dynamics
- Sea surface mechanism

- Radar system characteristics
- Target dynamics

The characterization of the sea surface, its dynamics, and its scattering are used to compute the radar cross section and Doppler spectrum of the backscattered signal. The quantities of interest are computed over a grid of discrete points in two dimensions of space, representing the surface, and time. The models require various environmental, radar, and geometric parameters as inputs.

10.2.1 Sea Surface Dynamics

The surface of the sea is characterized by evaluating its height, velocity, and slope over the two-dimensional surface and over time. The model needs as inputs the wind speed and its direction and depth of the water. As an intermediate step, the temporal and spatial correlations of the above quantities are computed in the form of frequency and wavenumber spectra. From this statistical description, sample functions of the above quantities are computed over the grid.

The following derivation is based on a one-dimensional surface to present the concept in a simplified notation. We start with the standard assumptions of incompressible and irrotational flow and define the potential function Φ_w such that

$$V = \nabla \Phi_w \tag{10.1}$$

where V is the velocity field of the fluid and ∇ is the gradient operator. Using the Bernoulli and continuity equations with suitable boundary conditions, the potential can be expressed as [Phillips, 1966]

$$\Phi_w = \sum_{n=1}^{\infty} \zeta^n (\Phi_w)_n \tag{10.2}$$

where ζ^n is a series expansion coefficient in units of surface slope.

Keeping only the term linear in ζ (i.e., the first-order Stokes wave), the solution for the potential, as a function of position and time, is

$$\Phi_{w1} = \frac{gA}{\omega} \frac{\cosh(k_w s)}{\cosh(k_w d)} \sin k_w (x - Ct) \tag{10.3}$$

where cosh = hyperbolic cosine

y = depth of water

d = wave travel in positive direction

$s = d + y$

x = horizontal direction of wave travel

k_w = wavenumber of fluid

A = wave amplitude

ω = wave frequency

C = phase velocity, ω/k_w

g = gravitational constant

From this potential, we can determine the velocity of a fluid element:

$$V_x = A_\omega \frac{\cosh k_w s}{\sinh k_w d} \cos k_w(x - Ct) \tag{10.4}$$

$$V_y = A_\omega \frac{\sinh k_w s}{\sinh k_w d} \sin k_w(x - Ct) \tag{10.5}$$

where sinh is the hyperbolic sine.

Thus it can be seen that when water is deep, the motion is circular. The radius of this circular motion is equal to the wave amplitude A. The surface height ζ is given by

$$\zeta = -\frac{1}{g} \frac{\partial \Phi_w}{\partial t}\bigg\|_{y=0} \tag{10.6}$$

$$= A \cos k_w(x - Ct)$$

The relation between the wavenumber and frequency of a wave is known as the dispersion relation. In its first-order form, it can be written as

$$\omega^2 = gk_w \left(1 + \frac{\gamma k_w^2}{g}\right) \tanh k_w d \tag{10.7}$$

where γ is the ratio of surface tension to water density and tanh is the hyperbolic tangent. When the above equations are generalized to a two-dimensional surface, as in the simulation implemented here, the wavenumbers became vector quantities.

The surface, in general, does not consist of just a single wave but can be represented by the superposition of many waves, covering a continuum of frequencies, fixed for one realization of the process. For a Gaussian surface, which is appropriate for a first-order Stokes approximation, the phases and amplitudes are drawn from uniform and Rayleigh distributions, respectively. A frequency spectrum (or wave height spectrum) $S(\omega)$ is required to determine the distribution of wave height versus frequency.

There exist a number of models that relate the enviromental parameters to the frequency spectrum. One such relation is the Pierson–Moskowitz frequency spectrum [Pierson and Moskowitz, 1964]. It represents the condition of a fully developed sea with unlimited fetch, as shown by

$$S(\omega) = 0.0081g^2\omega^{-5}\exp\left[0.74\left(\frac{\omega U_\omega}{g}\right)^{-4}\right] \tag{10.8}$$

where U_ω represents the wind speed. The wave frequency of the spectral peak is

$$\omega_p = 0.877\frac{g}{U_\omega} \tag{10.9}$$

The frequency spectrum can be extended to the directional frequency spectrum Ψ_0 by considering a directional distribution [Donelan and Pierson, 1987] such as

$$D(\omega, \theta) = \text{sech}^2[\beta(\omega)\theta] \tag{10.10}$$

where sech is the hyperbolic secant, and

$$\beta(\omega) = \begin{cases} 2.44\left(\dfrac{\omega}{0.95}\omega_p\right)^{+1.3} & \text{for } 0.56 < \dfrac{\omega}{\omega_p} < 0.95 \\[2mm] 2.44\left(\dfrac{\omega}{0.95}\omega_p\right)^{-1.3} & \text{for } 0.56 < \dfrac{\omega}{\omega_p} < 1.60 \\[2mm] 1.2 & \text{otherwise} \end{cases} \tag{10.11}$$

and

$$\Psi_0(\omega, \theta) = S(\omega)D(\omega, \theta) \tag{10.12}$$

The direction of the wind is taken as $\theta = 0$.

The directional wavenumber spectrum is derived from the directional frequency spectrum by using the relation

$$k_w dk_w = 2\frac{\omega^3}{g^2}d\omega \tag{10.13}$$

Given the wavenumber spectrum and the dispersion relation, a sample function of the surface is computed. This sample function is three dimensional: two spatial dimensions and time. The spectrum is made discrete over the wavenumber, then randomized, and converted to a surface by using a fast Fourier transform (FFT) subroutine. This procedure is repeated to compute the velocities and surface slopes. Figure 10.1 depicts a surface, at one instant of time, over a region of 1×1 km computed with a grid point spacing of 8×8 m.

Second-Order Stokes Waves

As the wave height increases, the nonlinear nature of the governing equations must be considered. One approach is to use a number of terms (two in this case) in the Stokes expansion. This is equivalent to considering the second-order nonlinear interaction between pairs of linear waves. In the case of deep water, the second-order surface term n_2 can be represented by

$$\eta_2(x, t) = \sum_{n=-N}^{N} \sum_{m=-N}^{N} \frac{F(n)F(m)}{2g} H(w_n, w_m; k_n, k_m)$$
$$\times \exp\{j[(w_n + w_m)t - (k_n + k_m)x]\} \quad (10.14)$$

where

$$H(w_n, w_m; k_n, k_m) = -g^2 \frac{k_n k_m}{w_n w_m} + (w_n + w_m)^2 - w_n w_m \quad (10.15)$$

and F is the (complex) amplitude spectrum of the surface sample function.

Computing the interaction between every pair of waves for a two-dimensional surface of useful size would be an extremely computational load. Alternatively, we may consider only self-interaction terms. The computational approach chosen in this work was to consider the interaction between all pairings of the N_i waves of highest amplitude where N_i was chosen as 1000 (i.e., 10^6 pairs).

Computing Surface Characteristics

The following parameters are computed from the amplitude spectrum of the sample function by implicit differentiation (if required) and two-dimensional FFT performed for each time sample:

- Surface height
- Surface Eulerian velocity (x, y, and z)
- Surface slope (in x and y directions)

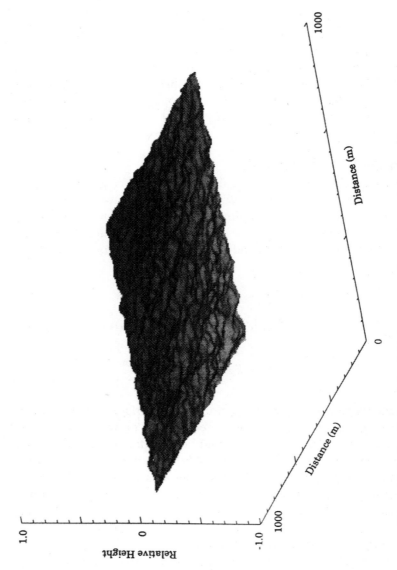

Figure 10.1 Example of simulated ocean surface at one instant of time based on theoretical model. Shown is surface in 1 × 1 km region using a grid point spacing of 8 × 8 m. Vertical dimension has been exaggerated to show wave better. Calculation of series of such surfaces can model temporal behavior of ocean surface.

These are computed on a spatial grid of up to 128×128 points with typically 8-m patches and up to 128 time steps with a typical resolution of 0.5 s. The spatial extent of the grid in both x and y should be at least as long as the longest wave. Several of these patches (typically 16) make up one radar resolution cell.

10.2.2 Sea Surface Scattering

For the purpose of the surface scatter model, two types of waves are identified. First, there are gravity waves, with wavelengths ranging from 200 m to a fraction of a meter. As pointed out in Chapter 2, the dominant restoring force for these waves is the force of gravity. Second, there are smaller capillary waves, with wavelengths on the order of centimeters or less. The dominant restoring force for these waves is surface tension.

The radar backscatter from the surface at each patch, defined by the gridded surface, is determined by using a two-scale composite model [Wright, 1968]. At microwave frequencies and low-to-medium grazing angles, the mechanism is Bragg scattering. The features of the surface with the same scale as half the radar wavelength are predominant scatterers. At microwave frequencies, the Bragg scattering is from the capillary waves. The two-scale composite scattering model considers the Bragg return as it is modulated by the larger scale gravity waves. Gravity waves influence the backscattering mechanism in the following ways:

- Surface tilting
- Advection of small scales due to orbital velocity
- Straining of small scales caused by acceleration (not included in this study)

The surface, considered first to be frozen in time, and the backscatter cross section, at points defined by the surface grid, are computed. Each surface grid point may be viewed as a plane tilted by the large-scale gravity waves. This plane is perturbed by capillary waves acting as Bragg scatterers. The backscatter cross section of such a plane is a function of the radar incidence angle as well as the local surface tilt both in the plane of propagation and normal to it. The backscattering mechanism also depends on the transmit and receive polarizations. The Bragg scattering from a tilted plane can be mathematically represented as

$$\sigma^{\circ}_{HH} = 16\pi |k_r|^4 \, \sin \psi^4 q^2(\psi, \delta)\Psi(k_{r\parallel}) \tag{10.16}$$

where

$$q^2(\psi, \delta) = \left| \left(\frac{\alpha \cos \delta}{\alpha_i} \right)^2 g_{HH}(\theta_i) + \left(\frac{\sin \delta}{\alpha_i} \right)^2 g_{VV}(\theta_i) \right|^2 \qquad (10.17)$$

where k_r is the radar wavenumber, $k_{r\parallel}$ is the radar wavenumber projected onto a local surface, δ is the surface tilt perpendicular to the plane of incidence, and θ_i is the local incidence angle. Similar expressions can be derived for σ°_{VV} and σ°_{HV}. This represents the long-term or modulation component of scattering, which is typically gamma distributed. It does not include the Rayleigh component. The exact relationship between the sea surface wavenumber and scattering geometry can be found in Donelan and Pierson [1987].

However, at low grazing angles, not every patch of the surface may be visible. Two different shadowing mechanisms are considered. One is to ignore those patches that are tilted away from the radar by more than 90°. The other is a statistical technique [Wetzel, 1970] that considers the probability of a point on the surface being obscured by a wave closer to the observer. For low grazing angles, we can compute a threshold height above which points on the surface are almost certain to be visible whereas those below it are not.

Using the above techniques, the backscatter cross section is computed for each patch on a surface with:

- Grazing angle, 10°
- Area preserved, 1 × 1 km on 8 × 8-m grid points
- Wind speed, 10 m/s
- Radar looking downwind

The spatial grid resolution effectively determines the cutoff between the two scales. The radar backscatter is represented by shaded surfaces in Figs. 10.2, 10.3, and 10.4 for the HH, VV, and HV polarizations, respectively. As noted in the measured results, the HH polarization is more spiky than the VV polarization.

These figures represent the spatial distribution of expected back-scattered power at one instant of time. Because of its nonlinear relation to surface slope, the backscattered power appears spiky. Whereas a radar with low resolution, which illuminates a larger surface area, would

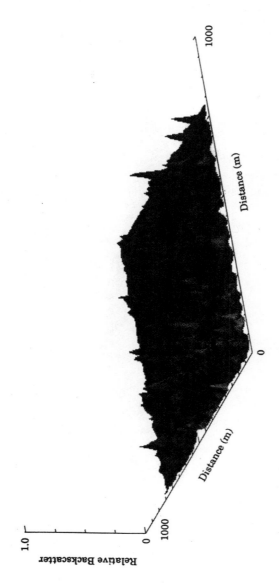

Figure 10.2 Example of simulated ocean surface at one instant of time for HH polarization based on theoretical model. Backscatter shown for surface in 1 × 1 m region using grid point spacing of 8 × 8 m. Vertical dimension has been exaggerated to show wave better.

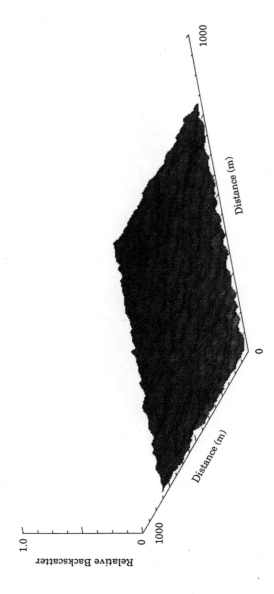

Figure 10.3 Example of simulated ocean surface at one instant of time for VV polarization based on theoretical model. Backscatter is shown for surface in 1 × 1 m region using grid point spacing of 8 × 8 m. Vertical dimension has been exaggerated to show wave better. Backscatter is not as spiky as HH polarization of Figure 10.2.

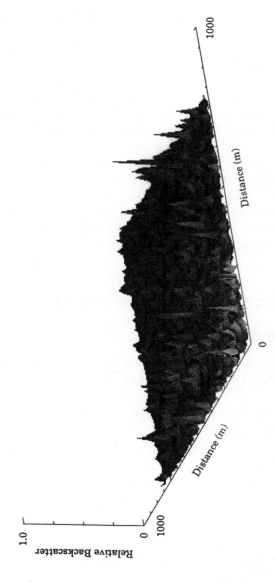

Figure 10.4 Example of simulated ocean surface at one instant of time for HV polarization based on theoretical model. Backscatter is shown for surface in 1 × 1 m region using grid point spacing of 8 × 8 m. Vertical dimension has been exaggerated to show wave better. Backscatter exhibits some spiky behavior caused by H component of polarization.

average out such spikes, one with higher resolution would not, leading to spiky temporal behavior.

10.2.3 Radar Model

The backscatter signal is represented in terms of "instantaneous" Doppler spectra computed at each grid time point. Two factors contribute to the above spectra from a given patch. They are (i) Doppler due to the orbital velocity of the patch (w_0) and (ii) frequency shift due to the Bragg scatter (w_b). Doppler shift due to Bragg scattering is the frequency of the wave with wavenumber $k_{r\parallel}$. There are two Doppler components: $w_0 + w_b$ and $w_0 - w_b$. The sample cross-section weighted mean and standard deviation of the Doppler shift over all patches in a given radar resolution cell is computed. It is calculated by fitting a Gaussian Doppler spectral shape to the power-weighted Doppler shifts of each patch that makes up the radar resolution cell.

10.3 COMPARISON CONDITIONS AND CRITERIA

10.3.1 Comparison Criteria

The criteria by which the measurement and simulation results are compared for the HH, VV, and HV polarizations are:

- Absolute cross section (dBsm/m/m)
- Amplitude distributions
- Estimated K-distribution shape parameter
- Mean Doppler (by pulse-pair method for measurements)
- Doppler width (by pulse-pair method for measurements)
- Temporal correlation of clutter power (HH)

10.3.2 Determining Measurement Conditions and Simulation Parameters

The following ground-truthing data were available for the measurements that were performed in Dartmouth, Nova Scotia:

- Wind speed and direction recorded every hour at Shearwater

- Wind speed and direction recorded sporadically at the IPIX radar
- Significant wave height and peak period of waves (also mean height and period) recorded at a buoy approximately 10 m offshore (and down shore) every few hours
- Where possible, directional wavenumber of peak of two-dimensional spectrum from radar scanning data

The high nonlinearity in the relation between surface slope and cross section (especially at low grazing angles) makes the retrieval of the true shape of the directional wavenumber spectrum essentially impossible without resorting to higher order spectral estimation techniques.

10.3.3 Comparison of Measurement and Simulation Results

Osborne Head Gunnery Range (OHGR) data sets collected on four different days are used to compare the model. The radar parameters for the measurements are as follows:

- Pulse width = 200 ns
- Radar height = 30 m above mean sea level (AMSL)
- No targets present in the data analyzed (clutter only)
- Alternate polarization on transmit, 2 kHz PRF

10.4 RESULTS OF CHAOTIC CHARACTERIZATION

The parameters of the simulated sea clutter data are listed in Table 10.1. Calculation of the chaotic invariants (correlation dimension and Lyapunov exponents) along with the nonlinearity tests were performed on the amplitude component of the simulated sea clutter data.

The amplitude of the simulated complex sea clutter data was computed for 50,000 samples. The time plot of this amplitude signal is shown in Fig. 10.5. The spectrum of the simulated data is shown in Fig. 10.6; its rate of decay with frequency is much more smaller than the spectrum of the FIR filtered sea clutter data shown in Fig. 5.4f.

The nonlinearity of this simulated sea clutter data was tested using the SIPD method where we made use of the surrogate data analysis with the growth of the interpoint distances as the discriminating statistic. This test is in accordance with the fact that chaotic systems are nonlinear by

Table 10.1   Measurement/Simulation Conditions and Parameter Values for Sea Clutter

Recorded Conditions	
Data and time	November 6, 1993
Wind speed	17 gusts to 32 km/h, 52 guests to 67 km/h at 8:45 AM
Wind direction	220° (with respect to north)
	170° (with respect to north) at 7:00 AM
Peak wave period	9.5 s
RMS wave height	1.2 m
Derived Wave Conditions	
Date and time	November 6, 1993, 2:31 PM
Peak wavelength	120 m
Peak wave direction	167° (with respect to north)
Measurement file	November 6/surv2
Measurement Parameters	
Radar range	2000 m
Azimuth	145° (with respect to north) 2:18 AM
Measurement file	November 6/starea5, 2:18 PM
Simulation Parameters	
Wind speed	47 km/h
Peak wavelength	114 m
Peak period	9.5 s
RMS wave height	0.91 m
Relative wave direction	22°
Water depth	20 m
Radar range	2000 m

Note: $N = 50,000$, 1-m grid resolution $= 128 \times 128$ spatial points surface evolves at a lag of 0.1 s, polarization VV.

definition. The average Mann-Whitney rank-sum statistic (Z-value) was shown to be -1.95, which implies that the simulated sea clutter data are produced by a linear process (a Z-value of less than -3.0 is considered to be a ground for strong rejection of the null hypothesis that the two observed processes are of a linear process). Chaotic invariants estimated for the simulated clutter data are presented in Table 10.2.

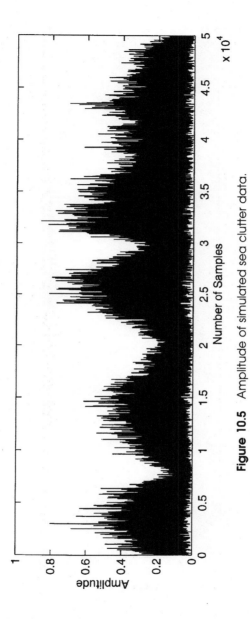

Figure 10.5 Amplitude of simulated sea clutter data.

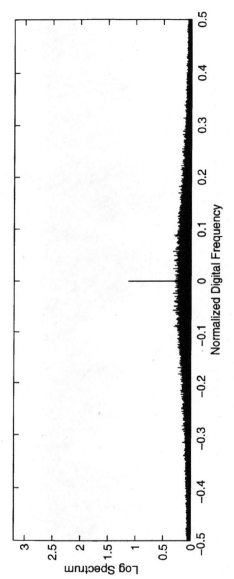

Figure 10.6 The Fourier spectrum of simulated sea clutter data.

Table 10.2   Chaotic Invariants for the Simulated Clutter Data Set

Parameter	˙ Amplitude of Simulated Clutter Data
τ	5
d_E	6
d_L	6
Z	-1.95
D_2	5.08
D_{KY}	5.85
$\lambda_i's$	$+0.5659$
	$+0.4123$
	$+0.2295$
	-0.0132
	-0.3279
	-1.0168

The embedding parameters (such as embedding dimension d_E and characteristic time delay τ) were determined using the global false nearest neighbor (GFNN) and the mutual information (MI) algorithms, respectively. Figure 10.7 shows the results obtained using the MI algorithm. The optimum time delay (τ) was found to be 5 (where the MI algorithm reaches a minimum for the first time). The optimum embedding dimension for the same data was found to be $d_E = 6$ from the GFNN analysis presented in Fig. 10.8. For the GFNN analysis, we used thresholds R_{tol1} and R_{tol2} of 25 and 10, respectively, and a time delay $\tau = 5$. The local false-nearest-neighbor (LFNN) analysis was performed on the simulated data set to find the local embedding dimension d_L (which gives the number of true Lyapunov exponents). Results of this analysis, presented in Fig. 10.9, show that a local embedding dimension $d_L = 6$ is required to estimate the Lyapunov exponents. The four curves shown in Fig. 10.9 correspond to local nearest neighbors (N_B) of 40, 60, 80, and 100. The results presented in Fig. 10.9 for the simulated sea clutter data exhibit a striking resemblance to those of Fig. 8.11 for colored noise, much more so than than those presented in Fig. 8.9 for real-life sea clutter data.

As mentioned earlier, the correlation dimension is the most basic and statistic property of an attractor. We used the STB_1 algorithm for the estimation of the correlation dimension of the simulated sea clutter data. The multidimensional embedding parameters were obtained using the

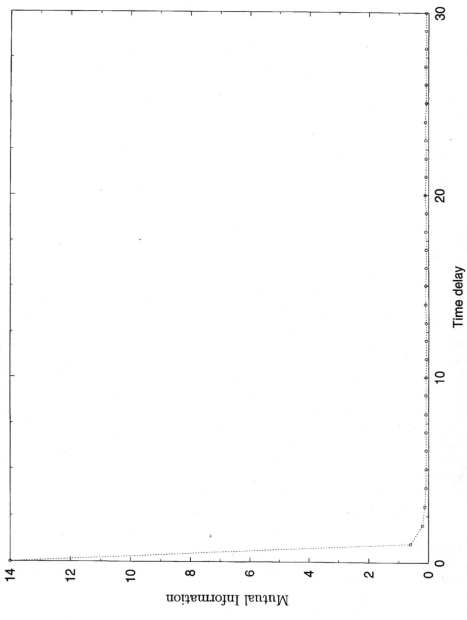

Figure 10.7 Mutual information for simulated sea clutter data.

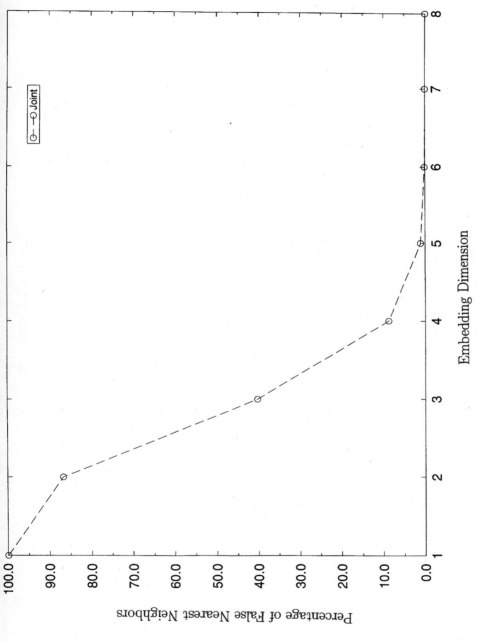

Figure 10.8 The GFNN for simulated sea clutter data.

185

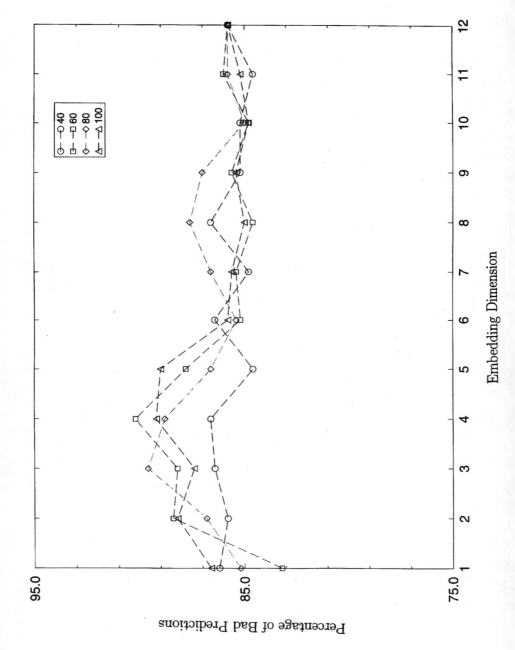

186

GFNN and AMI algorithms, respectively. The D_2 value for this particular data set was estimated to be 5.08 using the STB_1 algorithm.

Determination of the Lyapunov spectrum is particularly important in the analysis of a possible chaotic process. This is because the Lyapunov exponents not only tell us about the sensitive dependence (which is one of the fundamental characteristics of a true chaotic process) on the initial conditions of nearby trajectories, but they also tell us how orbits on the attractor move apart/together under the evolution of the dynamics. We used the BBA algorithm for estimating the Lyapunov exponents of the simulated sea clutter data. The algorithm gave six Lyapunov exponents corresponding to the embedding dimension used. A data length of 50,000 samples and a minimum of 3000 starting locations were used in the calculation. The analysis showed that the simulated data sets have three positive Lyapunov exponents with the fourth exponent close to zero followed by two negative exponents. The Lyupunov exponents for this case are shown in Fig. 10.10.

The Lyapunov dimension, or the Kaplan–Yorke dimension D_{KY}, for the simulated sea clutter data was estimated from the Lyapunov exponents. This is conjectured to be the same as that of the correlation dimension. The value of D_{KY} was estimated to be 5.86. The discrepancy between D_{KY} and the correlation dimension value, $D_{ML} = 5.08$, estimated using the STB_1 algorithm was, on the whole, much greater than that for actual sea clutter data.

The results of our study on characterizing the simulated sea clutter show that it satisfies chaotic requirements in some instances and violates them in other respects.

The summary of our findings is presented below.

Features Where the Simulated Sea Clutter Agrees with a Chaotic Model

1. The simulated sea clutter does not behave like noise. This is based on the results which we obtained from the GFNN analysis, which clearly produced an embedding dimension of 6. For the noise data GFNN never gives an optimum embedding dimension (see Fig. 8.10).

2. The sum of all Lyapunov exponents is negative, suggesting that the simulated clutter is a dissipative process.

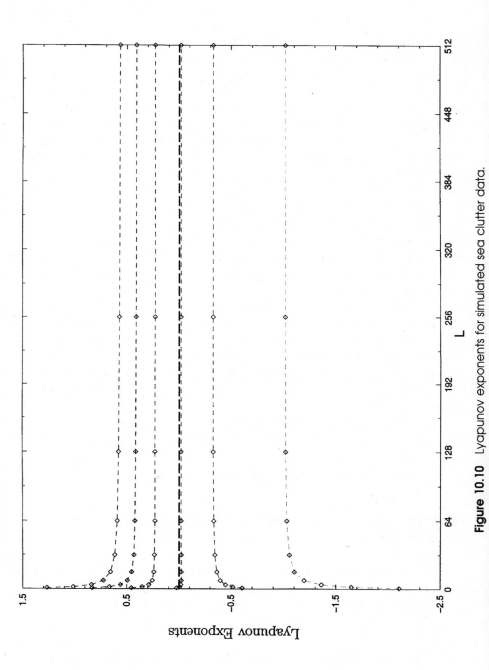

Figure 10.10 Lyapunov exponents for simulated sea clutter data.

Shortcomings of Simulated Sea Clutter

1. The spectrum of the simulated clutter is relatively flat (see Fig. 10.6), which is unlike the $1/f$ spectrum measured for real-life sea clutter (see Fig. 5.4).
2. The simulated clutter failed the nonlinearity test, a necessary requirement for a chaotic process.
3. The LFNN analysis for simulated clutter (Fig. 10.9) is different from that of real-life sea clutter (Fig. 8.9).
4. The Lyapunov spectrum for the simulated clutter is more complex with three or more positive Lyapunov exponents.
5. The difference between D_{ML} and D_{KY} for the simulated clutter is larger than in the case of real-life sea clutter data.

Concluding Remarks

There are two important observations that can be made from this study.

1. It is misleading to make statements on the chaotic nature of a process based on one or more parameters such as the correlation dimension and largest Lyapunov exponent. The only correct way in which a process can be said to be chaotic is for all the following conditions to be satisfied:
   a. The process is nonlinear.
   b. The spectrum follows a $1/f$ law.
   c. The correlation dimension is fractal.
   d. At least one Lyapunov exponent is positive.
   e. The sum of all Lyapunov exponents is negative.
   f. There is close agreement between the correlation dimension and the Kaplan–Yorke dimension.
   g. The local dimension is less than or equal to the global embedding dimension, that is, $d_L < d_E$.

   This is indeed how it is for real-life sea clutter data [Haykin and Puthusserypady, 1997], as demonstrated in the previous chapter.
2. The chaotic dynamics of real-life sea clutter provide a framework against which the physical validity of other models of sea clutter could be tested.

11

SUMMARY OF EXPERIMENTAL RESULTS AND CONCLUSIONS

The experimental study reported in this book has clearly demonstrated the chaotic dynamics of sea clutter in the most complete manner. This conclusion is based on the results of different real-life radar experiments, data that were used to compute the correlation dimension, Lyapunov exponents, and Kaplan–Yorke dimension. The results are summarized here:

- The correlation dimension D_2 of real-life sea clutter using a maximum likelihood procedure lies in the range 4.1–4.5, independently of time of the day, month of the year, radar range and range resolution, type of like polarization, sea state, radar location, and radar type. It is also essentially the same regardless of whether the radar component used in the computation is the in-phase, quadrature-phase, or amplitude component.
- The global embedding dimension d_E is essentially 5.
- The local embedding dimension (dynamic embedding dimension) d_L is essentially 5. This shows that the number of true Lyapunov exponents is 5, with $d_L = d_E$.
- The Lyapunov spectrum consists of two positive exponents followed by one exponent equal to zero (within experimental errors) and two negative exponents. This composition of the Lyapunov spectrum is

191

independent of both radar range and the radar component being considered. Moreover, the distribution of the Lyapunov exponents is independent of sea state. However, their absolute values do depend on sea state and radar type.

- The Kaplan–Yorke dimension, a by-product of the Lyapunov spectrum, is consistently very close to D_2.

- The Kolmogorov entropy computed using a maximum-likelihood procedure is fairly close to the sum of positive Lyapunov exponents, another by-product of the Lyapunov spectrum.

These results were obtained using carefully ground-truthed real-life radar data collected with an instrument-quality radar at two different locations and a commercial marine radar at another location. The data were calibrated as appropriate and preprocessed to minimize the effects of noise and roundoff errors, always taking extra care to protect the integrity of the actual clutter process.

Leung and Haykin [1990] presented preliminary results obtained using the Grassberger–Procaccia algorithm in light of which we posed the question:

Is there a radar clutter attractor?

On the basis of the many positive results presented in this book, we are now emboldened to answer this question with an emphatic "yes."

Furthermore, the various parameter estimation algorithms used in this book have revealed highly significant information about the chaotic dynamics of sea clutter, as concluded here:

1. Sea clutter, as observed by a microwave radar (regardless of whether it is coherent or noncoherent) at low grazing angles, is generated by a low-degree-of-freedom strange attractor. The small value of correlation dimension D_2, lying between 4.1 and 4.5, for varying environmental conditions is especially interesting since the intrinsic dissipation is relatively weak.

2. The results of Lyapunov spectrum analysis reveal that the generation of sea clutter is governed by a coupled system of five nonlinear differential equations: This is a remarkable result deduced from the fact that the third Lyanpunov exponent is consistently zero (within

numerical errors) for an embedding dimension of 5. However, the derivation of these equations requires that we go back to the basic physical principles that govern the generation of sea clutter. This is indeed a difficult research topic, which is most deserving of detailed attention.

3. The correlation dimension and Kaplan–Yorke dimension are relatively insensitive to variations in sea state. In contrast, the absolute values of Lyapunov exponents, for a given range and radar system, do exhibit significant variations with respect to sea state. This latter result demonstrates how the behavior of a chaotic attractor can vary within a fixed dimensionality. The result is also reassuring as it emphasizes the nonstationary nature of sea clutter observed over a long period of time. It should, however, be noted that for each sea state the Lyapunov spectrum analysis was performed using a set of sea clutter data short enough to satisfy stationarity of the data as required by the estimation algorithms and yet long enough to produce reliable results.

11.1 SUPPORTING PHYSICAL EVIDENCE

There is also strong supporting physical evidence for our research findings, as indicated in what follows.

Sea clutter is naturally related to fluid turbulence in an ocean environment. Ruelle and Takens [1971] proposed a chaotic mechanism for the generation of turbulence in a fluid and related phenomena in a dissipative system. The existence of chaos in fluid turbulence was subsequently confirmed by Abarbanel et al. [1994]. Indeed, the low value of D_2 or D_{KY} in the range of 4.1–4.5 reported here for sea clutter is typical for fluid-related phenomena.

In Chapter 2, we discussed the role of gravity waves in the generation of sea clutter. In this context, the results of chaotic analysis performed by Frison and Abarbanel [1997] on ocean gravity waves measured at the Harvest platform off the coast of California are highly relevant. They reported a $D_{KY} \approx 4.5$, which is essentially the same as that for sea clutter reported herein. The value of D_{KY} reported by Frison and Abarbanel is computed from a Lyapunov spectrum for ocean gravity waves that is identical in structure to that of sea clutter.

11.2 DYNAMIC RECONSTRUCTION

The following quote, taken from the preface to a book by Gray and Davisson [1986] on random processes, is ever so true in the case of sea clutter:

> One can argue that given complete knowledge of the physics of an experiment, the outcome must always be predictable, at least with the aid of a sufficiently powerful computer.

Although we do not have knowledge of the system of nonlinear differential equations that describes the physical generation of sea clutter, the research findings reported in this book point to the existence of such a system consisting of five equations. This observation, in turn, encourages us to invoke two important theorems:

1. **Delay Embedding Theorems** [Takens, 1981]: The underlying nonlinear dynamics of a physical (chaotic) phenomenon can be reconstructed in the form of a predictive model by using a time series based on one variable of the phenomenon.
2. **Universal Approximation Theorem** [Haykin, 1999]: Given data representative of a physical (chaotic) phenomenon, it is possible to construct a neural network (in the form of a multilayer perceptron or radial-basis function network) that approximates the predictive model.

Thus, despite the lack of knowledge of the exact physical laws describing the generation of sea clutter, we are assured by these two theorems that we can indeed build a predictive model capable of capturing the underlying nonlinear dynamics of sea clutter. Preliminary results on construction of this predictive model in the form of a regularized radial-basis function network for sea clutter under varying conditions are described elsewhere [Haykin et al., 1998]. The results presented therein provide further proof for the chaotic dynamics of sea clutter.

APPENDIX

In this appendix, we illustrate the results of chaotic invariant analysis that were performed on Lorentz data. These results are presented to show the power and accuracy of all the algorithms that we used for the estimation of invariants of a well-known chaotic process.

The Lorenz attractor is described by the following system of equations:

$$\dot{x} = \sigma(y - x)$$
$$\dot{y} = -xz + rx - y \qquad \text{(A.1)}$$
$$\dot{z} = xy - bz$$

where $x, y,$ and z are all functions of time and $\dot{x}, \dot{y},$ and $\dot{z}$ are their derivatives with respect to time.

We used the standard parameter values $\sigma = 16$, $b = 4$, and $r = 45.92$ for generating the $x, y,$ and z time series of the Lorentz attractor.

For the above set of parameters, the theoretical values for the fractal (correlation) dimension D is 2.06. There are three Lyapunov exponents and their values are $\lambda_1 = 1.50$, $\lambda_2 = 0.0$, and $\lambda_3 = -22.5$. The Kolmogorov entropy is 1.50.

A.1 ESTIMATION OF CHAOTIC INVARIANTS FROM MEASUREMENTS OF A SINGLE VARIABLE

In this section we discuss the results on chaotic invariant analysis performed on the x component of the Lorenz map described above. The first thing we have to compute is the embedding (or generating a pseudo-embedding space). This computation needs the estimation of the embedding parameters: the embedding delay (τ) and global embedding dimension (d_E). We applied the methods described in Chapter 8 for the

195

estimation of these embedding parameters to reconstruct the state space. Once the state space is constructed, we applied the different algorithms described in Chapter 9 for the estimation of the invariants of the system.

A.1.1   Results

Figure A.1 plots the mutual information versus time delay, producing an optimum delay equal to 4. The embedding dimension estimate was performed using the method of false nearest neighbors, yielding the value of 3, as shown in Fig. A.2. The local embedding dimension d_L was estimated using the method of local false nearest neighbors, obtaining the value of 3, as shown in Fig. A.3. The Lyapunov spectrum was estimated using the BBA method, yielding the results shown in Fig. A.4. The estimated values are compared to the theoretical values in Table A.1. It can be seen that these two sets of values are very close to each other. The correlation dimension was estimated using the STB_1 algorithm, yielding the value of 2.06. This value is very close to the theoretical value of the correlation dimension, as is clear from Table A.1. The Kaplan–Yorke dimensions estimated using the theoretical and experimental Lyapunov spectra are very close to each other. Similarly, the Kolmogorov entropy estimated using the STB_2 algorithm and the corresponding values computed from the Lyapunov spectra are also in close agreement. The experimental results presented in this appendix for the Lorenz attractor confirm the reliability of the algorithms used in this book for estimating the invariants of a chaotic process.

Table A.1   Comparison of Chaotic Invariants of Lopez Attractor

Time Series	d_E	τ	D_2	D_{KY}	Lyapunov Exponents (nats/s)		
					λ_1	λ_2	λ_3
Theoretical	3	—	2.06	2.067	+1.50	0.0	−22.50
Reconstructed	3	4	2.07	2.069	+1.5697	−0.0314	−22.31

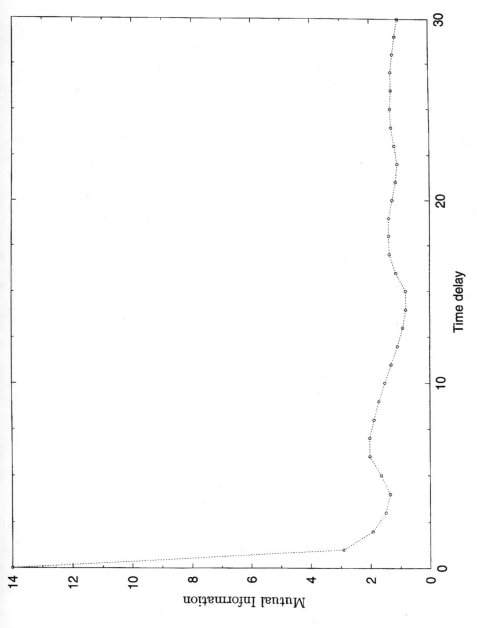

Figure A.1 Mutual information for Lorenz data.

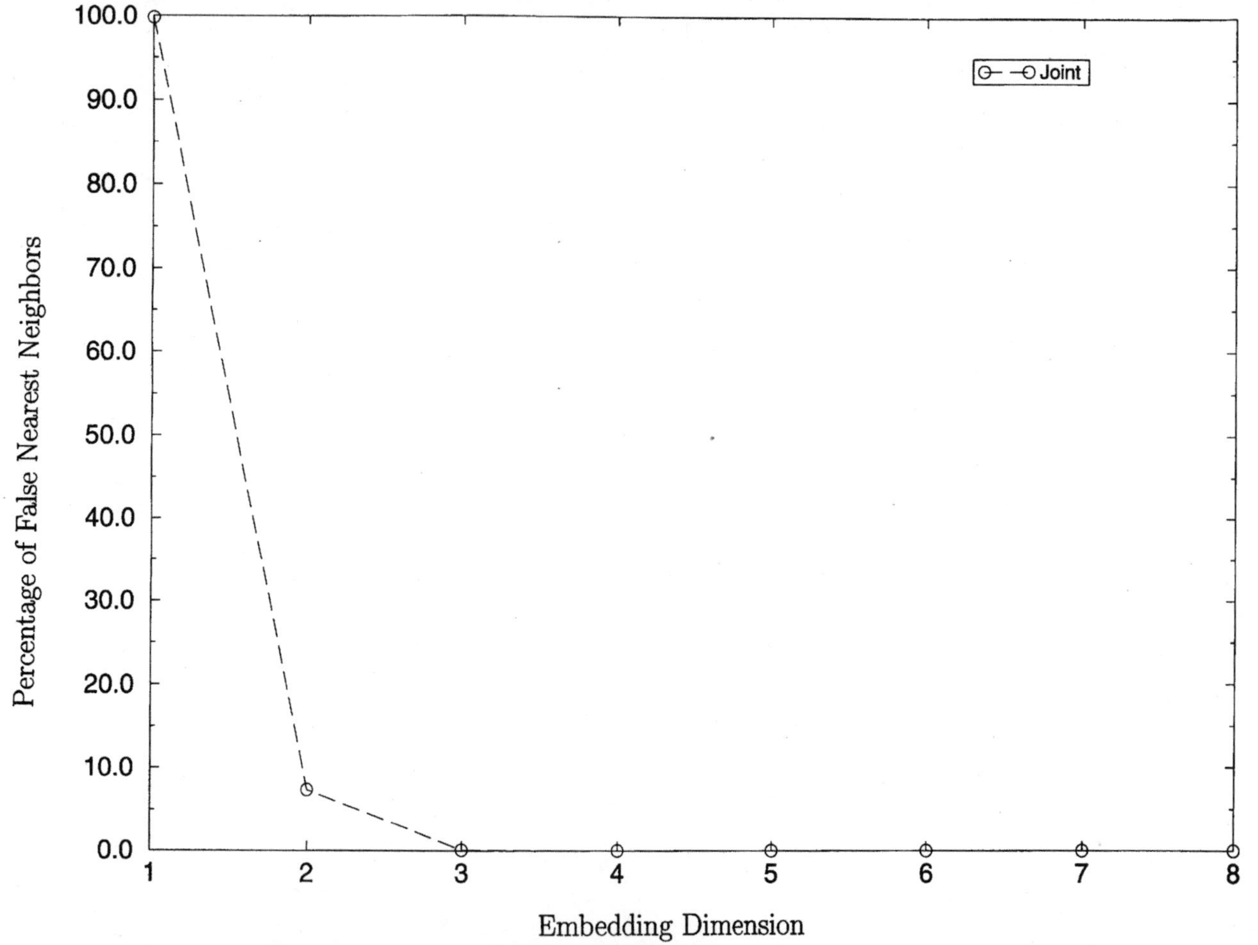

Figure A.2 The GFNN for Lorenz data.

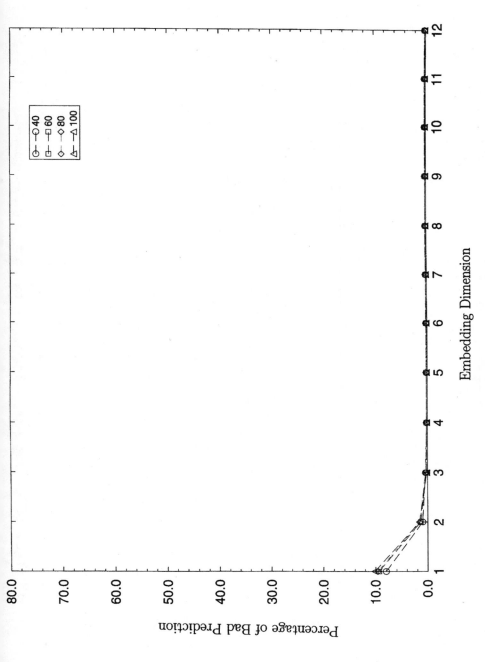

Figure A.3 LFNN for Lorenz data.

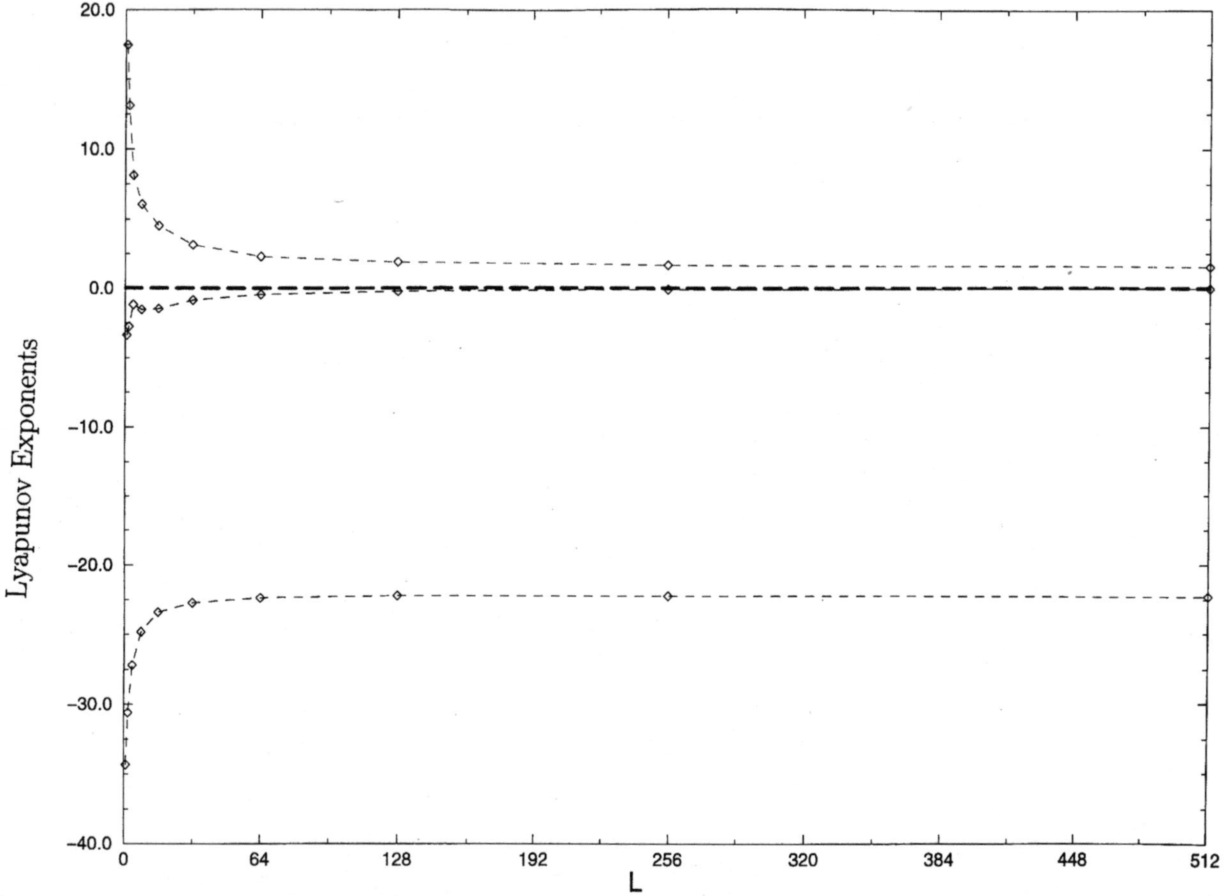

Figure A.4 Estimated Lyapunov exponents of Lorenz data using BBA algorithm.

BIBLIOGRAPHY

Abarbanel, H. D. I., *Analysis of Observed Chaotic Data*, Springer-Verlag, New York, 1996.

Abarbanel, H. D. I., and Kennel, M. B., "Local False Nearest Neighbors and Dynamical Dimensions from Observed Chaotic Data," *Physical Review E* **47**, 3057–3068 (1993).

Abarbanel, H. D. I., and Sushchik, M. M., "True Lyapunov Exponents and Models of Chaotic Data," *International Journal of Bifurcation and Chaos* **3**, 543–550 (1993a).

Abarbanel, H. D. I., and Sushchik, M. M., "Local or Dynamical Dimensions of Nonlinear Systems Inferred from Observations," *International Journal of Bifurcation and Chaos* **3**, 543–550 (1993b).

Abarbanel, H. D. I., Brown, R., and Kennel, M. B., "Variation of Lyapunov Exponents on a Strange Attractor," *Journal of Nonlinear Science* **1**, 175–199 (1991a).

Abarbanel, H. D. I., Brown, R., and Kennel, M. B., "Lyapunov Exponents in Chaotic Systems: Their Importance and Their Evaluation Using Observed Data," *International Journal of Modern Physics* **B5**, 1347–1375 (1991b).

Abarbanel, H. D. I., Brown, R., and Kennel, M. B., "Local Lyapunov Exponents Computed from Observed Data," *Journal of Nonlinear Science* **2**, 343–365 (1992).

Abarbanel, H. D. I., Brown, R., Sidorowich, J. J., Tsimring, L. S., "The Analysis of Observed Chaotic Data in Physical Systems," *Review of Modern Physics* **65**, 1331–1392 (1993).

Abarbanel, H. D. I., Katz, R., Cembrola, J., Galeb, T., and Frison, T., "Nonlinear Analysis of High Reynold Number Flows over a Buoyant Asymmetric Body," *Physical Review E* **49**, 4003–4018 (1994).

Abarbanel, H. D. I., Frison, T. W., and Tsimring, Lev. Sh., "Obtaining Order in a World of Chaos," *IEEE Signal Processing Magazine* **15**, 49–65 (1998).

Apel, J. R., *Principles of Ocean Physics*, International Geophysics Series, 38, Academic Press, New York, 1987.

Arnold, S. F., The Theory of Linear Models and Multivariate Analysis, Wiley, New York, 1981.

Badii, R., Broggi, G., Derighetti, B., and Ravani, M., "Dimension Increases in Filtered Chaotic Signals," *Physical Review Letters* **60**, 979–982 (1988).

Bak, P., Tang, C., and Wiesenfeld, K., "Self-organized Criticality: An Explanation of $1/f$ Noise," *Physical Review Letters* **59**, 381–388 (1987).

Baker, G. L., and Gollub, J. P., *Chaotic Dynamics: An Introduction*, 2nd ed., Cambridge University Press, Cambridge, 1996.

Balatoni, J., and Renyi, A., Publications of the Mathematical Institute of the Hungarian Academy of Science **1**, 9–13, 1956. English translation, Selected Papers of A. Renyi, **1**, 558–562, 1976. See also Renyi, A., *Acta Mathematica* **10**, 193–197 (1959).

Barna, G., and Tsuda, A. I., "New Method for Computing Lyapunov Exponents," *Physics Letters A* **175**, 421–427 (1993).

Beckmann, P., and Spizzichino, A., "*The Scattering of Electromagnetic Waves from Rough Surfaces*," Pergamon, Oxford, 1963.

Benettin, G., Galgani, L., Giorgilli, A., and Strelcyn, J. M., "Tous Les Nombres Characteristiques de Lyapunov Sont Effectivement Calculables," *Comptes Rendues Academi of Science, Paris* **286A**, 431–433 (1978).

Bennettin, G., Galgani, L., Giorgilli, A., and Strelcyn, J. M., "Lyapunov Characteristic Exponents for Smooth Dynamical Systems and for Hamiltonian Systems: A Method for Computing All of Them. Part 1: Theory," *Meccanica* **15**, 9–20 (1980a).

Benettin, G., Galgani, L., Giorgilli, A., and Strelcyn, J. M., "Lyapunov Characteristic Exponents for Smooth Dynamical Systems and for Hamiltonian Systems: A Method for Computing All of Them. Part 2: Numerical Application," *Meccanica* **15**, 21–30 (1980b).

Billingsley, P., *Ergodic Theory and Information*, Wiley, New York, 1965.

Booker, H. G., Ratcliffe, J. A., and Shinn, D. H., "Diffraction from an Irregular Screen with Applications to Ionospheric Problems," *Philosophical Transcations of Royal Society of London A* **242**, 579–607 (1950).

Briggs, K., "An Improved Method for Estimating Lyapunov Exponents of Chaotic Time Series," *Physics Letters A* **151**, 27–32 (1990).

Broomhead, D. S., Huke, J. P., and Muldoon, M. R., "Linear Filters and Nonlinear Systems," *Journal of Royal Statistical Society* **54**, 373–382 (1992).

Broomhead, D. S., and King, G. P., "Extracting Quantitative Dynamics from Experimental Data," *Physica D* **20**, 217–226 (1986).

Brown, R., Bryant, P., and Abarbanel, H. D. I., "Computing the Lyapunov Spectrum of a Dynamical System from Observed Time Series," *Physical Review A* **43**, 2787–2806 (1991).

Bryant, P., Brown, R., and Abarbanel, H. D. I., "Lyapunov Exponents from Observed Time Series," *Physical Review Letters* **65**, 1523–1526 (1990).

Clarke, A. B., and Disney, R. L., *Probability and Random Processes for Engineers and Scientists*, Wiley, New York, 1970.

Dieci, L., and Van Vleck, E., "Computation of a Few Lyapunov Exponents for Continuous and Discrete Dynamical Systems," *Applied Numerical Mathematics* **17**, 275–291 (1995).

Crutchfield, J. P., Farmer, J. D., Packard, N. H., and Shaw, R. S., "Chaos," *Scientific American*, **255**, 46–57 (1986).

Diks, C., "Estimating Invariants of Noisy Attractor," *Physical Review E* **53**, R4263–R4266 (1996).

Ding, M., Grebogi, C., and Ott, E., "Evolution of Attractors in Quasiperiodically Forced Systems: From Quasiperiodic to Strange Nonchaotic to Chaotic," *Physics Review A*, **39**, 2593–2598, (1989).

Ding, M., Grebogi, C., Ott, E., Sauer, T., and Yorke, J. A., "Estimating Correlation Dimensions from a Chaotic Time Series: When Does Plateau Onset Occur?," *Physica D* **69**, 404–424 (1993).

Donelan, M. A., and Pierson, Jr., W. J., "Radar Scattering and Equilibrium Ranges in Wind Generated Waves with Applications to Scatterometry," *Journal of Geophysical Research* **92**(5), 4971–5029 (1987).

Drosopoulos, T., private communication, DREO, Ontario, Canada, July 1997.

Eckmann, J. P., Kamphorst, S. O., and Ruelle, D., "Recurrence Plots of Dynamical Systems," *Europhysics Letters* **4**, 973–977 (1987).

Eckmann, J. P., Kamphorst, S. O., Ruelle, D., and Ciliberto, S., "Lyapunov Exponents from Time Series," *Physical Review A* **34**, 4971–4979 (1986).

Eckmann, J. P., and Ruelle, D., "Ergodic Theory of Chaotic and Strange Attractors," *Review of Modern Physics* **57**, 617–656 (1985).

Eckmann, J. P., and Ruelle, D., "Fundamental Limitations for Estimating Dimension and Lyapunov Exponents in Dynamical Systems," *Physica D* **56**, 185–187 (1992).

Ellner, S., "Estimating Attractor Dimensions from Limited Data: A New Method, with Error Estimates," *Physics Letters A* **133**, 128–133 (1988).

Essex, C., and Narenberg, M. A. H., "Comments on Deterministic Chaos: The Science and the Fiction by Ruelle," *Proceedings of The Royal Society of London A* **435**, 287–292 (1991).

Farmer, J. D., "Information Dimension and the Probabilistic Structure of Chaos," *Zeitschrift für Naturforsch* **37a**, 1304–1314 (1982a).

Farmer, J. D., "Dimension, Fractal Measure and Chaotic Dynamics," in H. Haken (Ed.), *Evolution of Order and Chaos*, Springer, Heidelberg, New York, 1982b.

Farmer, J. D., "Sensitive Dependence on Parameters in Nonlinear Dynamics," *Physical Review Letters* **55**, 351–354 (1985).

Farmer, J. D., Ott, E., and Yorke, J. A., "The Dimension of Chaotic Attractors," *Physica D* **7**, 153–180 (1983).

Fay, F. S., Clarke, J., and Peters, R. S., "Weibull Distribution Applied to Sea Clutter," *Proceedings of IEE Conference (Radar '77)*, 101–103 (1977).

Fraser, A. M., "Information and Entropy in Strange Attractors," *IEEE Transactions on Information Theory* **35**, 245–262 (1989).

Fraser, A. M., and Swinney, H. L., "Independent Co-ordinates for Strange Attractors from Mutual Information," *Physical Review A* **33**, 1134–1140 (1986).

Frison, T. W., and Abarbanel, H. D. I., "Ocean Gravity Waves: A Nonlinear Analysis of Observations," *Journal of Geophysical Research* **102**, 1051–1059 (1997).

Fukunga, K., and Olsen, D. R., "An Algorithm for Finding the Instrinsic Dimensionality of Data," *IEEE Transactions on Computers* **20**, 176–183 (1971).

Gallager, R. G., *Information Theory and Reliable Communication*, Wiley, New York, 1968.

Geist, K., Parlitz, U., and Lauterborn, W., "Comparison of Different Methods for Computing Lyapunov Exponents," *Progress in Theoretical Physics* **85**, 875–893 (1990).

Goldstein, H., "Sea Echo in Propagation of Short Radio Waves," *MIT Radiation Lab. Ser.*, **13**, set 6.6, D. E. Kerr, Ed., McGraw-Hill, New York, 1951.

Gollub, G. H., and Van Loan, C. F., *Matrix Computations* 2nd Ed. John Hopkins, Baltimore, 1989.

Grassberger, P., in *Chaos, Nonlinear Science: Theory and Applications*, A. V. Holden, Ed., Manchester University Press, Manchester, 1987.

Grassberger, P., "An Optimized Box-Assisted Algorithm for Fractal Dimension," *Physics Letters A* **148**, 63–68 (1990).

Grassberger, P., and Procaccia, I., "Characterization of Strange Attractors," *Physical Review Letters* **50**, 346–351 (1983a).

Grassberger, P., and Procaccia, I., "Measuring the Strangeness of Strange Attractors," *Physica D* **9**, 189–208 (1983b).

Gray, R. M., and Davisson, L. D., *Random Processes: A Mathematical Approach for Engineers*, Prentice-Hall, Englewood Cliffs, NJ, 1986.

Grebogi, C., Ott, E., Pelikan, S., and Yorke, J. A., "Strange Attractors that are not Chaotic," *Physica D*, **13**, 261–268 (1984).

Greene, J. M., and Kim, J. S., "The Calculation of Lyapunov Spectrum," *Physica D* **24**, 213–225 (1987).

Guckenheimer, J., and Holmes, P. J., *Nonlinear Oscillations, Dynamical Systems, and Bifurcations of Vector Fields*, Springer-Verlag, New York, 1983.

Hamburger, D., Krasnor, C., Haykin, S., Currie, B. W., and Nohara, T. J., "A Computer-Integrated Radar System for Research Use," *Proceedings of the Canadian Conference on Electrical and Computer Engineering*, (Montreal, Canada), 594–596 (1989).

Haykin, S., "Chaotic Signal Processing: New Research Directions and Novel Applications," *Proceedings of IEEE Workshop on SSAP*, Victoria, BC, Oct. 1992.

Haykin, S., *Neural Networks, A Comprehensive Foundation*, Second Edition, Prentice-Hall, Upper Saddle River, 1999.

Haykin, S., "Chaotic Characterization of Sea Clutter: New Experimental Results and Novel Applications," *Proceedings of the 29th Asilomar Conference on Signals, Systems and Computers*, Monterey, CA, October (1995).

Haykin, S., and Leung, H., "Chaotic Model of Sea Clutter Using a Neural Network," *Proceedings of the SPIE Conference*, San Diego, CA. 1989.

Haykin, S., and Leung, H., "Model Reconstruction of Chaotic Dynamics: First Preliminary Radar Results," *Proceedings of the International Conference on Acoustics Speech and Signal Processing* **IV**, 125–128 (1992).

Haykin, S., Krasnor, C., Nohara, T. J., Currie, B. W., and Hamburger, D., "A Coherent Dual-Polarized Radar for Studying the Ocean Environment," *IEEE Transactions on Geoscience and Remote Sensing* **29**, 189–191 (1991).

Haykin, S., Kezys, V., and Currie, B., "Surface-Based Radar: Coherent," in *Remote Sensing of Sea Ice and Icebergs*, S. Haykin, E. O. Lewis, R. K. Raney, and J. R. Rossiter, Eds., 443–504 Wiley, New York (1994).

Haykin, S., Krasnor, C., Nohara, T. J., Currie, B. W., and Hamburger, D., "A Coherent Dual-Polarized Radar for Studying the Ocean Environment," *IEEE Transactions on Geoscience and Remote Sensing*, **29**, 189–191 (1991).

Haykin, S., and Li, X. B., "Detection of Signals in Chaos," *Proceedings of IEEE* **83**, 94–122 (1995).

Haykin, S., and Puthusserypady, S., "Chaotic Dynamics of Sea Clutter," *Chaos* **7**, 777–802 (1997).

Haykin, S., Puthusserypady, S., and Yee, P., "Dynamic Reconstruction of Sea Clutter Using Regularized RBF Networks," in *Proceedings of the 32nd Asilomar Conference on Signals, Systems and Computers*, Pacific Grove, CA, Nov. 1–4, 1998.

He, N. and Haykin, S., "Chaotic Modelling of Sea Clutter," *Electronics Letters* **28**, 2076–2077 (1992).

Hediger, T., Passamante, A., and Farrell, M. E., "Characterizing Attractors Using Local Intrinsic Dimensions Calculated by Using SVD and Information Theoretric Criteria," *Physical Review A* **41**, 5325–5332 (1990).

Henderson, H. W., and Wells, R., "Obtaining Attractor Dimensions from Metereological Time series," *Advances in Geophysics* **30**, 205–237 (1988).

Hénon, M., "A Two Dimensional Mapping with a Strange Attractor," *Communications in Mathematical Physics* **50**, 69–77 (1976).

Jakeman, E., and Pusey, P. N., "A Model for Non-Rayleigh Sea Echo," *IEEE Transactions on Antennas and Propagation* **24**, 806–814 (1976).

Jakeman, E., and Tough, R. J. A., "Non-Gaussian Models for the Statistics of Scattered Waves," *Advances in Physics* **37**, 471–529 (1988).

Kaplan, J., and Yorke, J., "Chaotic Behavior of Multidimensional Difference Equations," *Lecture Notes in Mathematics* **730**, 228–237 (1979).

Kaplan, J., Yorke, E, and Yorke, J., "The Lyapunov Dimension of Strange Attractors," *Journal of Differential Equations* **49**, 185–207 (1983).

Kennel, M. B., Brown, R., and Abarbanel, H. D. I., "Determining Embedding Dimension for Phase-space Reconstruction Using a Geometrical Construction," *Physical Review E* **45**, 3403–3411 (1992).

Kolmogorov, A. N., "A New Metric Invariant of Transitive Dynamical Systems and Automorphisms in Lebesgue Spaces," *Doklady Akademii Nauk SSSR* **119**, 861–864 (1958).

Krasnor, C., Haykin, S., Nohara, T. J., and Currie, B. W., "A New coherent Marine Radar for Ocean-Related Studies," *Proceedings of the Canadian Conference on Electrical and Computer Engineering*, Vancouver, Canada, 711–714 (1988).

Krasnor, C., Haykin, S., Currie, B. W., and Nohara, T. J., "A Coherent Dual-Polarized Radar for Ice Surveillance Studies," *International Conference on Radar*, Paris, 438–443 (1989).

Landa, P. S., and Rosenblum, M. G., "Time Series Analysis for System Identification and Diagnostics," *Physica D*, **48**, 232–254 (1991).

F. Ledrappier, "Some Relations between Dimension and Lyapunov Exponents," *Communications in Mathematical Physics* **81**, 229–238 (1981).

Leung, H., and Haykin, S., "Is There a Radar Clutter Attractor," *Applied Physics Letters* **56**, 593–595 (1990).

Li, X. B., "Detection of Signals in Chaos," Ph.D. thesis, McMaster University, Canada, 1995.

Li, X. B., and Haykin, S., "Chaotic Detection of Small Targets in Sea Clutter," *Proceedings of the International Conference on Acoustics Speech and Signal Processing* **1**, 237–240 (1993).

Li, X. B., and Haykin, S., "Chaotic Characterization of Sea Clutter," l'Onde Electrique, Special Issue on Radar, SEE, France, 60–65 (1995).

Liebert, W., and Schuster, H. G., "Proper Choice of the Time-delay for the Analysis of Chaotic Time Series," *Physics Letters A*, **142**, 107–111 (1989).

Liebert, W., Pawelzik, K., and Schuster, H. G., "Optical Embedding of Chaotic Attractors from Topological Considerations," *Europhysics Letters* **14**, 521–526 (1991).

Longuett-Higgins, M. S., "The Distribution of Intervals between Zeros of a Stationary Random Function," *Philosophical Transactions of the Royal Society of London A* **254**, 557–599 (1962).

Lorenz, E.N., "Determinstic Non-periodic Flow," *Journal of Atmospheric Science* **20**, 130–141 (1963).

Mandelbrot, B. B., *Fractals: Form, Chance and Dimension*, Freeman, San Francisco, 1977.

Mandelbrot, B. B., *Fractal Geometry of Nature*, Freeman, New York, 1983.

Mañé, R., "On the Dimension of Compact Invariant Sets of Certain Nonlinear Maps," *Lecture Notes in Mathematics* **898**, 230–242 (1981).

May, R. M., "Simple Mathematical Models with Very Complicated Dynamics," *Nature*, **261**, 459–467 (1976).

Mayer-Kress, G., and Hubler, A., "Time Evolution of Local Complexity Measures and Aperiodic Perturbations of Nonlinear Dynamical Systems," in *Measures of Complexity and Chaos*, N. B. Abraham et al., Eds., Plenum Press, New York, 1986.

Mercier, R. P., "Diffraction by a Screen Causing Large Random Phase Fluctuations," *Proceedings of Cambridge Phylosophical Society A* **58**, 382–400 (1962).

Mitschke, F., "Acausal Filters for Chaotic Signals," *Physical Review A*, **41** 1169–1171 (1990).

Moon, F. C., *Chaotic and Fractal Dynamics: An Introduction for Applied Scientists and Engineers*, Wiley, New York, 1992.

Nathanson, F. E., *Radar Design Principles: Signal Processing and the Environment*, McGraw-Hill, New York, 1969.

Nohara, T. J., and Haykin, S., "Canadian East Coast Radar Trials and the K-Distribution," *IEE Proceedings, Part F* **138**, 80–88 (1991).

Olofsen, E., Degoede, J., and Heijungs, R., "A Maximum Likelihood Approach to Correlation Dimension and Entropy Estimation, *Bulletin of Mathematics and Biology*, **54**, 45–58 (1992).

Ornstein, D. S., *Ergodic Theory, Randomness and Dynamical Systems*, Yale University, New Haven, CT, 1974.

Osborne, A. R., and Provenzale, A., "Finite Correlation Dimension for Stochastic Systems with Power-Law Spectra,"*Physica D* **35**, 357–381 (1989).

Oseledec, V. I., "A Multiplicative Ergodic Theorem. Lyapunov Characteristic Numbers for Dynamical Systems," *Trudy Mosk. Mat. Obsc. Moscow Math. Soc.* **19**, 17–28 (1968).

Ott, E., *Chaos in Dynamical Systems*, Cambridge University Press, New York, 1993.

Packard, N. H., Crutchfield, J. P., Farmer, J. D., and Shaw, R. S., "Geometry from a Time Series," *Physical Review Letters* **45**, 712–716 (1980).

Paladin, G., and Vulpiani, A., "Anomalous Scaling Laws in Multi-fractal Objects," *Physics Reports* **156**, 147–225 (1987).

Palmer, A. J., Kropfli, R. A., and Fairall, C. W., "Signature of Deterministic Chaos in Radar Sea Clutter and Ocean Surface Winds," *Chaos* **5**, 613–616 (1995).

Paoli, P., Politi, A., Broggi, G., Ravani, M., and Badii, R., "Phase Transitions in Filtered Chaotic Signals," *Physical Review Letters* **62**, 2429–2432 (1989).

Parlitz, U., "Identification of True and Spurious Lyapunov Exponents from Time Series," *International Journal of Bifurcation and Chaos* **2**, 155–165 (1992).

Pecora, L. M., and Carrol, T. L., "Discontinuous and Non-differentiable Functions and Dimension Increase, Induced by Filtering Chaotic Data," *Chaos* **6**, 432–439 (1996).

Pecora, L. M., Carrol, T. L., and Heagy, J. F., "Statistics for Mathematical Properties of Maps between Time Series Embeddings," *Physical Review E* **52**, 3420–3439 (1995).

Peitgen, H. O., Jurgens, H., and Saupe, D., *Chaos and Fractals*, Springer-Verlag, New York, 1992.

Pesin, Ya. B., "Characteristic Lyapunov Exponents and Smooth Ergodic Theory," *Uspeki Matematicheskikh Nauk* **32**(4), 55–71 (1977).

Phillips, O. M. *The Dynamics of the Upper Ocean*, Cambridge University Press, London, 1966.

Pierson, W. J., and Moskowitz, L., "A Proposed Spectral Form for Fully Developed Wind Seas Based on the Similarity Theory of S. A. Kitaigorod-skii," *Journal of Geophysical Research* **69**(24), 5181–5203 (1964).

Pineda, F. J., and Sommerer, J. C., "A Fast Algorithm for Estimating the Generalized Dimension and Choosing Time Delays," in *Time Series Prediction: Forecasting the Future and Understanding the Past*, A. S. Weigend and N. A. Gershenfeld, Eds., 1994, pp. 367–385.

Poincaré, H., *Les Methodes Nouvelles de la Mécanique Celeste*, vols 1–3, Gauthier-Villars, Paris, 1892, 1893, 1894. (Translation: *New Methods of Celestial Mechanics*, NASA, 1967).

Priestley, M. B., *Nonlinear and Nonstationary Time Series*, Academic, New York, 1988.

Provenzale, A., Smith, L. A., Vio, R., and Murante, G., "Distinguishing between low-dimensional Dynamics and Randomness in Measured Time Series," *Physica D* **58**, 31–49 (1992).

Rapp, P. E., Albano, A. M., Schmah, T. I., and Farwell, L. A., "Filtered Noise Can Mimic Low-Dimensional Chaotic Attractors," *Physical Review E* **47**, 2289–2297 (1993).

Rice, S. O., "The Mathematical Analysis of Random Noise," *Bell Systems Technical Journal* **23**, 282–332 (1944).

Rice, S. O., "The Mathematical Analysis of Random Noise," *Bell Systems Technical Journal* **24**, 46–156 (1945).

Rössler, O. E., "An Equation for Continuous Chaos," *Physics Letters A* **57**, 397–398 (1976).

Romeiras, F. J., Bondeson, A., Ott, E., Edwar, E., Antonsen, Th. M. Jr., and Grebogi, C., "Quasiperiodically Forced Dynamical Systems with Strange Nonchaotic Attractors," *Physica D*, **26**, 277–294 (1987).

Romeiras, F. J., Bondeson, A., Ott, E., Antonsen, Th. M. Jr., and Grebogi, C., "Quasiperiodic forcing and the observability of strange non-chaotic attractors," *Physica Scripta*, **40**, 442 (1989).

Ruelle, D., *Elements of Differentiable Dynamics and Bifurcation Theory*, Academic Press, New York, 1989.

Ruelle, D., "Ergodic theory of differentiable dynamic systems," *Publications in Physics and Mathematics, IHES*, **50**, 275–306 (1979).

Ruelle, D., "Deterministic Chaos: The Science and the Fiction," *Proceedings of the Royal Society of London* **A-427**, 241–248 (1990).

Ruelle, D., and Takens, F., "On the Nature of Turbulence," *Communications in Mathematical Physics* **20**, 167–192 (1971).

Russell, D. A., Hanson, J. E., and Ott, E., "Dimensionality and Lyapunov Numbers of Strange Attractors," *Physical Review Letters*, **45**, 1175–1178 (1980).

Salpeter, E. E., "Interplanetary Scintillations: I. Theory," *Astrophysical Journal* **147**, 433–448 (1967).

Sano, M., and Sawada, Y., "Measurement of the Lyapunov Exponents from a Chaotic Time Series," *Physical Review Letters* **55**, 1082–1085 (1985).

Sauer, T., and Yorke, J. A., "How Many Delay Coordinates Do You Need?," *International Journal of Bifurcation and Chaos*, **3**, 737–744 (1993).

Sauer, T., Yorke, J. A., and Casdagli, M., "Embedology," *Journal of Statistical Physics* **65**, 579–617 (1991).

Schouten, J. C., Takens, F., Van den Bleek, C. M., *RRCHAOS: A Menu-Driven Software Package for Chaotic Time Series Analysis*, Section Chemical Reactor Engineering, Delft University of Technology, Delft, Netherlands, 1994a.

Schouten, J. C., Takens, F., and Van den Bleek, C. M., "Estimation of Dimension of a Noisy Attractor," *Physical Review E* **50**, 1851–1861 (1994a).

Schouten, J. C., Takens, F., and van den Bleek, C. M., "Maximum-likelihood Estimation of the Entropy of an Attractor," *Physical Review E*, **49**, 126–129 (1994b).

Schuster, H. G., *Deterministic Chaos*, VCH, Verlagsgessellschaft, mbH Weinheim, Germany, 1988.

Seber, G. A. F., *Multivariate Observations*, Wiley, New York, 1984.

Shimada, I., and Nagashima, T., "A Numerical Approach to Ergodic Problem of Dissipative Dynamical Systems," *Progress in Theoretical Physics* **61**, 1605–1611 (1979).

Sidorowich, J. J., "Modelling of Chaotic Time Series for Prediction, Interpolation and Smoothing," *Proceedings of International Conference on Acoustics Speech and Signal Processing* **4**, 121–124 (1992).

Sigel, S., *Non-parametric Statistics for the Behavioral Sciences*, International Student Edition, McGraw-Hill, Japan, 1956.

Sinai, Ya. G., "On the Concept of Entropy of a Dynamical System," *Doklady Akademii Nauk SSSR* **124**, 768–771 (1959).

Sinai, YA. G., *Introduction to Ergodic Theory*, Princeton University Press. Princeton, NJ, 1976.

Shannon, C. E., and Weaver, W., *The Mathematical Theory of Communication*, University of Illinois Press, Urbana, 1949.

Skolnik, M. I., *Introduction to Radar Systems*, 2nd ed., McGraw-Hill, New York, 1980.

Smale, S., "Differentiable Dynamical Systems," *Bulletin of the American Mathematical Society* **73**, 747–817 (1967).

Stoop, R., and Meier, P., "Evaluation of Lyapunov Exponents and Scaling Functions from Time Series," *Journal of Optical Society of America* **B-5**, 1037–1045 (1988).

Stoop, R., and Parisi, J., "Calculation of Lyapunov Exponents Avoiding Spurious Elements," *Physica D* **50**, 89–94 (1991).

Takens F. "Detecting Strange Attractors in Turbulence," *Lecture Notes in Mathematics* **898**, 366–381 (1981).

Takens, F., "Invariants Related to Dimension and Entropy," *Atas do 13° Coloqkio Brasileiro de Matematica*, Instituto de Matematica Pure e Aplicada, Rio de Janeiro, 1983a.

Takens, F. "On the Numerical Determination of the Dimension of an Attractor," *Lecture Notes in Mathematics* **1125**, 99–106 (1983b).

Theiler, J., "Quantifying Chaos: Practical Estimation of the Correlation Dimension" Ph.D. Thesis, Caltech, 1988.

Theiler, J., "Some Comments on the Correlation Dimension of $1/f^{\alpha}$ Noise," *Physics Letters A* **155**, 480–493 (1990a).

Theiler, J., "Estimating Fractal Dimension," *Journal of the Optical Society of America* **A7**, 1055–1073 (1990b).

Theiler, J., Eubank, S., Longtin, A., Galdrikian, B., and Farmer, J. D., "Testing for Nonlinearity in Time Series: The Method of Surrogate Data," *Physica D* **58**, 77–94 (1992).

Thomson, J. M. T., and Stewart, H. B., *Nonlinear Dynamics and Chaos: Geometrical Methods for Engineers and Scientists*, Wiley, New York, 1986.

Tong, H., *Nonlinear Time Series Analysis: A Dynamical System Approach*, Oxford University Press, Oxford, 1990.

Tools for Dynamics [Chaotic Signal Processing (CSP) software], Applied Chaos, LLC, Del Mar, CA, 1995.

Trunk, G. V., "Radar Properties of Non-Rayleigh Sea Clutter," *IEEE Transactions on Aerospace and Electronics Systems* **8**, 196–204 (1972).

von Breman, H. F., Udwadia, F. E., and Proskurowski, W., "An efficient QR Based Method for the Computation of Lyapunov Exponents," *Physica D* **101**, 1–16 (1997).

Walpole, R. E., and Myers, R. H., *Probability and Statistics for Engineers and Scientists*, Macmillan, New York, 1972.

Wetzel, L. B., "Sea Clutter", in *Radar Handbook*, M. I. Skolnik, Ed., McGraw-Hill, New York, 1970, pp. 13.1–13.40.

Whitney, H., "Differentiable Manifolds," *Annals of Mathematics* **37**, 645–680 (1936).

Williams, R. F., "The DA Maps of Smale and Structural Stability," *Proceedings of the Symposium of Pure Mathematics* **14**, 324–334 (1970).

Wolf, A., Swift, J. B., Swinney, H. L., and Vastano, J. A., "Determining Lyapunov Exponents from a Time Series" *Physica D* **16**, 285–317 (1985).

Wright, J. W., "A New Model for Sea Clutter," *IEEE Transactions on Antennas and Propagation* **16**, 217–223 (1968).

Young, L. S., "Dimension, Entropy and Lyapunov Exponents," *Journal of Ergodic Theory and Dynamical Systems* **2**, 109–114 (1982).

Young, L. S., "Entropy Lyapunov Exponents, and Hausdorff Dimensions in Differentiable Dynamical Systems" *IEEE Transactions on Circuits and Systems* **30**, 599–607 (1983).

INDEX

A/D converter, 47, 72
Air–sea interactions, 9
Algorithms, *see specific types of algorithms*
Amplitude:
    corrections, 48–52
    fluctuations in, generally, 12, 48–52
    Kolmogorov entropy (KE) and, 163
    simulation model, 179
Angles, low-grazing, 9–11
Antennas, footprint of, 7–8
Attractor(s), generally:
    in chaotic dynamics, 28–31
    in dynamic systems, 26
    phase-space, 105
    reconstruction, 109
    strange, *see* Strange attractor
Autocorrelation function, 102, 108–109

Backscatter:
    determining   factors, 7–8
    signal, 173, 178
Basin of attraction, 31
BBA algorithm, 4, 133, 143, 148, 187
Box-counting dimension (D_0), 26, 34–35
Bragg scattering, 10–12, 173, 178
Broadband, 31
Built-in calibration equipment (BICE),
    16–17

Calibration:
    IPIX radar, 16–17
    I-Q, 48–52
Capacity dimension, 26
Capillary waves, 9–11
Chaos, defined, 79–80
Chaos theory, 2

Chaotic dynamics:
    assessment criteria, 46
    attractors, 28–31
    basin of attraction, 31
    fractal dimension, 32–34
    fractal geometry and, 28
    historical perspective, 25–26
    limit cycle, 32
    manifold, 32
    sink, 32
    strange attractor, 32
Chaotic processes:
    nonlinear dynamical systems, 85–88
    overview, 79–80
    phase space, defined, 88–89
    purpose of, generally, 89–90
    sensitive dependence on initial
        conditions, 81–85
C^n topology, 96, 98
Computer control, IPIX radar, 17
Continuity:
    defined, 59
    test of, 60–61, 64–66
Continuous-time dynamical system, 87, 101
Correlated noise, 114, 118, 120
Correlation dimension:
    defined, 26, 35
    estimation methods, 4, 121–130
    nonlinearity tests, 75
    simulation model, 193
Correlation function:
    on attractor, generally, 34
    differentiable embedding, 105
Covariance matrix, local embedding
    dimension, 112–113

D_2:
  defined, 3, 191
  estimation methods, 121–127
Data acquisition system, IPIX radar, 16
Delay coordinate function:
  differentiable embedding, 105
  topological embedding, 100–103
Delay embedding theorems, 194
Determinism, Laplacian, 81
Deterministic system, defined, 87
Diffeomorphism, 58
Differentiability:
  defined, 61–62
  test of, 62–66
Differentiable embedding, 103–106
Discrete-time dynamical system, 87
Divergence/convergence, Lyapunov
  exponents, 36
Doppler spectra, 178
Drift signal, 68–69
Dual polarization, IPIX radar, 15
Dynamical dimension, 107
Dynamic reconstruction, 194

Embedding:
  chaotic characterization, 183
  defined, 25, 41
  delay, see Embedding delay
  dimension, see Embedding dimension
Embedding delay:
  autocorrelation function, 109
  defined, 45
  implications of, generally, 108
  mutual information method, 109–110
  normalized, 4, 106
Embedding dimension:
  defined, 45
  phase-space reconstruction, 106
  stationarity test, 68, 71
Embedding space, reconstruction of:
  differentiable embedding, 103–106
  delay coordinates, 100–103
  embedding parameters, 108–120
  topological embedding, 93–100
Embedding theory, 91–93, 129
Entropy:
  Kolmogorov (KE), 27, 43–45, 76
  metric, 46, 156
Estimation:
  of correlation dimension, 121–130
  horizon of predictability (HP), 147–151
  Kolmogorov entropy, 156–165

Lyapunov dimension, 151–156
Lyapunov exponents, 130–147
  from single variable, 195–200
Euclidean norm, 158
Evolutionary processes, 87–88

Filtered noise, 118
Filtering:
  FIR, 54–56, 127–128, 143
  three-point smoothing, 52–54, 145
Finite impulse response (FIR) filtering:
  characteristics of, 54–56
  defined, 52
  estimation methods and, 127–128, 143
  nonlinearity tests, 72
Flow, defined, 87
Fluid motion, 82
Fluid turbulence, 193
Footprint, antenna, 7, 9
Fourier transform (FFT), 57, 73, 171
Fractal dimension, 26–27, 32–34
Fractal geometry, 28
Frequency modulation (FM), 5

Gaussian noise theory, 11
Geometric probability density function, 159
GFNN algorithm:
  defined, 4
  embedding space, reconstruction of, 107,
    110, 113–114, 118, 120
  estimation methods, 127
  simulation model, 183, 187
Global embedding dimension:
  defined, 4
  estimation methods and, 110–111
  in nonlinearity tests, 76
  simulation model, 191
Global false nearest neighbor (GFNN),
  see GFNN algorithm
GPA, see Grassberger-Procaccia algorithm
  (GPA)
Grassberger-Procaccia algorithm (GPA), 3,
  122–124, 192
Gravity waves, 9–10, 193

Heaviside function, 33, 122
Henon attractor, 30
Hidden signal, 68–69
Hooke's laws, 88
Horizon of predictability (HP), 147–151

I-channels, 48–50

I-Q calibration:
  miscalibration, effect of, 50–52
  overview, 48–50
Immersion, 103
Information dimension (D_1), 26, 35
Information theory, 157
Initial conditions, 81–85, 157
In-phase components:
  correlation dimension, estimation of, 129
  Kolmogorov entropy, estimation of, 163
  Lyapunov exponents, estimation of, 143, 145–146
  in preprocessing of radar data, 48
Interference, constructive, 10
IPIX radar:
  built-in calibration, 16–17
  coherent transmission/reception, 15
  computer control, 17
  digital data acquisition, 16
  dual polarization, 15
  flexible operation and modification, 17
  pulse compression, 15–16
  system features and capabilities, overview, 14–15
  X-band transmission, 16

Jacobian matrices, 133–134, 143

Kaplan-Yorke dimension, 143, 187, 192–193. See also Lyapunov (Kaplan-Yorke) dimension
K-distributions, compound, 11–12
Kolmogorov entropy (KE):
  characteristics of, 43–44
  defined, 27
  estimation methods, 4, 157–165
  Lyapunov exponents, relationship between, 45
  maximum-likelihood estimation of, 159–165

Landau, Lev D., 82
Laplace, Pierre Simon de, 81
Least squares, differentiability index, 62
LFNN algorithm, 4, 108, 114, 189
Limit cycle, defined, 32
Lipschitz, 105
Local embedding dimension:
  defined, 4
  embedding space, reconstruction of, 111–113, 191
  simulation model, 183

Local false-nearest neighbor (LFNN), see LFNN algorithm
Logistic equation, 83
Lorenz, E. N., 29
Lorenz attractor, 29–30
Low-grazing angles, 9–11
Lyapunov (Kaplan-Yorke) dimension:
  defined, 26–27
  estimation methods, 151–156
  in preprocessing radar data, 41–43
  simulation model, 187, 192–193
Lyapunov exponent:
  defined, 3, 27
  differentiable embedding and, 105–106
  estimation methods, 130–147, 183
  Kolmogorov entropy (KE), relationship between, 45
  miscalibration and, 50
  nonlinearity tests, 75–76
  simulation model, 192–193
Lyapunov spectrum:
  calculation of, 38–41
  characteristics of, generally, 35–38
  simulation model, 189, 191–193

Manifold, defined, 32
Mann-Whitney rank-sum statistic (Z-value), 73, 75, 180
Map, defined, 32
Mathematical models, 88
Maximum-likelihood estimate, correlation dimension, 124–126
Maximum-likelihood principle, 4
Maximum norm, 158
May, Robert, 83
Meteorology, 29, 85
Metric entropy, 46, 156. See also Kolmogorov entropy (KE)
Microwave radar, 10–11, 173, 192
Miscalibration, 50–52
Mutual information (MI) algorithm:
  defined, 4
  embedding space, reconstruction of, 102, 113–114
  phase-space reconstruction, 107, 109
  simulation model, 183

Nats per sample, 37, 133
Nats per second, 37–38
Nearest neighbor, embedding delay, 107–112
Newton's laws, 88

Noise, generally:
 correlated, 114, 118, 120
 filtered, 118
 Gaussian theory, 11
 maximum-likelihood estimate,
  correlation dimension, 124–126
 process, nonlinearity tests, 74–75
Nonlinear dynamical systems, 2, 25, 85–89
Nonlinearity, in simulation model, 179. See
 also Nonlinear dynamical systems
Nonlinearity tests:
 results of, 74–78
 SCD method, 73–74
 SIPD method, 73
 WSF method, 72, 74
Nonstationary time series analysis, 68
Normalized embedding delay, 4, 106
Null hypothesis:
 continuity tests, 60–61
 differentiability tests, 63
 nonlinearity test, 73
 simulation model, 180

Orbits, estimation methods and, 156–157
Osborne Head Gunnery Range (OHGR),
 179

Passband response, 56
Periodicities, drift and hidden, 68–69
Phase space:
 defined, 88–89
 reconstruction, 106–108
Poincare, Henri, 82–83
Polarization:
 correlation dimension, 1, 191
 IPIX radar, 15
Population growth, 83
Predictability, 146. See also Horizon of
 predictability (HP)
Predictions, 80–81, 85, 87–88
Prevalence, 96, 98
Probability density function, 159
Pulse compression, IPIX radar, 15–16

Q-channels, 48–50
QR decomposition, 132
Quadrature-phase components, 48, 129,
 145–146, 163
Quantum mechanics, 83

Radar:
 data, see Radar data, preprocessing of

defined, 13
field experiments, 17–24
IPIX, 14–17
microwave, 10–11, 173, 192
model, 178
parameters, 9
Radar data, preprocessing of:
 amplitude and phase corrections (I-Q
  calibrations), 48–52
 continuity, 58–61, 64–66
 differentiality, 58, 61–66
 filtered data, 58
 filtering, 52–56
 results of, generally, 56–58
Radioactive decay, 81
Random fluid motion, 82
Randomness, 12
Random phenomena, 80–81
Random system, defined, 87
Random-walk model, 11
Rayleigh distributions, 170
Recurrence plot, stationarity test, 69
Resolution cells, 9
Rossler attractor, 30
Ruelle, D., 96

SCD algorithm, 73–75
SCD method, defined, 4
Sea, gravity waves, 9–10
Sea clutter, generally:
 defined, 2–3, 7
 at low-grazing angles, 9–11
 simulation model, 167–189
Sea echo, 7
Sea surface, simulation model:
 dynamics, 168–173
 scattering, 173–178
Second-order phase transition, 89
Sensitive dependence on initial conditions,
 81–85
Sensitivity time control (STC), 17
Signal-to-clutter ratio, 12
Signal-to-noise ratio (SNR), 3, 127, 132
Simulation model:
 chaotic characterization, 179–190
 comparison conditions, 178–179
 overview, 167–168
 sea surface dynamics, 168–173
 sea surface scattering, 173–178
 shortcomings of, 189
Sink, defined, 32
SIPD algorithm, 4, 73, 75, 179

State space, 87
Stationarity test, recurrence plots, 68–71
STB$_1$ algorithm, 4, 126–127
STB$_2$ algorithm, 4, 162–163
Stochastic system, defined, 87
Stokes approximation, 170
Stokes waves, 168, 171
Stopband response, 56
Strange attractors:
  defined, 29, 32
  fractal dimensions, 26, 33
  low-degree-of-freedom, 192
Surface acoustic wave, (SAW), 15
Surrogate data analysis, 76–77, 118
Swell, defined, 10

Takens's theorem, 25
Taylor series, 133
Temporal, defined, 7
Three-point smoothing, 52–54, 114, 145
Tides data, 120
Time series, nolinearity test, 4

Topological embedding, 93–100
Transistor-transistor-logic (TTL), 17
Transmission, IPIX radar, 15–16

Universal approximation theorem, 194
Unpredictability, 80–81, 85, 131

Vector(s), generally:
  in dynamical system, 87
  local embedding dimension, 113
  phase-space reconstruction, 106
Vector field, 87

Wave-shaping filter (WSF), 72, 74
WGN sequence, 70
Williams, R. F., 29
Wind speed and direction, 178–179

X-band transmission, IPIX radar, 16

Z-value, 73, 75, 180